D0772531

OUR NATIVE BEES

OUR NATIVE BEES

NORTH AMERICA'S ENDANGERED POLLINATORS AND THE FIGHT TO SAVE THEM

PAIGE EMBRY

Timber Press | Portland, Oregon

Photo and illustration credits appear on page 218.

Published in 2018 by Timber Press, Inc.
The Haseltine Building
133 S.W. Second Avenue, Suite 450
Portland, Oregon 97204-3527
timberpress.com

Printed in China
Text design by Adrianna Sutton
Jacket design by Kimberly Glyder and Adrianna Sutton

Catalog records for this book are available from the Library of Congress and the British Library.

ISBN 13: 978-1-60469-769-8

For Franklin's bumble bee and all the other bees that
are gone or at risk—even the ones we don't know about yet

Contents

Bees' appearance varies hugely. Some are massively hairy like this male *Habropoda excellens*, the three-spotted digger bee, from Utah.

What a Bee Is

An Introduction

NATIVE BEES ARE the poor stepchildren of the bee world. Honey bees get all the press—the books, the movie deals—and they aren't even from around here, coming over from Europe with the early colonists. In 2015, when President Barack Obama's White House issued a plan to restore 7 million acres of land for pollinators and more than double the research budget for them, it was called the National Strategy to Promote the Health of Honey Bees and Other Pollinators. Four thousand species of native bees, not to mention certain birds, bats, flies, wasps, beetles, moths, and butterflies, reduced to "other pollinators." Sigh.

Honey bees are fine bees. They dance and make honey and can be carted around by the thousands in convenient boxes, but from a pollination point of view, they aren't super-bees. On cool, cloudy days when honey bees are home shivering in their hives, many of our native bees are out working over the flowers. Bumble bees do their special buzz pollination of tomatoes, blueberries, and various wild species. Squash bees wake up early to catch the big yellow squash blossoms while they're open. The trusty orchard mason bees are such hard-working yet slovenly little pollen collectors that several hundred can pollinate an acre of apples that requires thousands of honey bees. Where are the book and movie deals for these bees?

Well, I'm making a small start for them here.

This female sweat bee, *Halictus ligatus*, has gotten herself smothered in pollen.

Native versus naturalized bees

A North American native bee is one that evolved right here. The honey bees we know, *Apis mellifera*, are not native because they came over from Europe with the early colonists. Some of those early bees quickly escaped into the wild (they went feral), where they did quite well. Those feral bees are considered naturalized, not native. Once upon a time, however, another member of the honey bee clan did live here. A fossil of *Apis nearctica* was found in Nevada in rocks laid down about 14 million years ago. It's the first member of the honey bee genus that's been found as a native in the New World.

It began with tomatoes

My obsession with bees began because of tomatoes, a plant with roots deep in my Georgia childhood. The summertime table in my house always had a plate of sliced tomatoes on it. My dad grew tomatoes wherever he could: along a brick wall at one house, in the only sunny spot by some azaleas at another, in a tiny plot of red dirt outside his office door for a while. When I grew up and moved away, I, too, grew tomatoes, although with varying success in the cool summers of Seattle. Tomatoes have been a fixture in my life.

Now, tomatoes have some flexibility in their pollination requirements. Some pollination happens as a result of wind just shaking the plants, but more and bigger tomatoes result with the help of bees. Not just any bee can do it, though. It wasn't until I was nearly fifty that I learned that honey bees can't produce those tasty red and orange globes. Tomatoes require a special kind of pollination called buzz pollination, where a bee holds onto a flower and vibrates certain muscles that shake the pollen right out of the plant. Honey bees don't know how to do it, but certain native bees do. I was appalled. How could I, a serious gardener for many years, not have learned that it takes a native bee—not a European import—to properly pollinate a tomato?

So I asked other people I knew, veteran gardeners and non-gardeners alike, and it turns out that I was not the slow-witted exception. Not only did the people I talked to not know that honey bees couldn't pollinate tomatoes, many didn't know that honey bees weren't native to North America. None of them knew that just in the United States and Canada there are 4000 species of bees that *are* native. Upon reflection, it's not that surprising. We mostly notice the troublesome things insects do: sting, eat the wood in our houses, chew up our plants. Truthfully, many of us don't even view insects as animals, although what else could they be? Plants? Fungus? Bacteria? No, insects are animals, and many of them do good things for us, but those good things creep by unnoticed.

I often see the statistic that one bite out of every three we take is thanks to pollinators, but every bite isn't created equal, either from a taste point of view or a nutritional one. Some of the foods that we eat the most, like wheat, corn, and rice, are either self-pollinated or wind-pollinated, but many of our most delicious fruits and vegetables are bee pollinated: strawberries, blueberries, apples, peaches, and, of course, tomatoes. Also, a good chunk of our essential nutrients are concentrated in animal-pollinated fruits and vegetables. Ninety-eight percent of the vitamin C, seventy percent of the vitamin A, fifty-five percent of the folic acid, and seventy-four percent of the lipids come from animal-pollinated plants. In a 2013 *Scientific American* article, University of California Berkeley conservation biologist Claire Kremen said that if the pollinators all died off, we might not starve to death, but we'd likely get some sort of vitamin deficiency disorder.

One can also put at least part of what pollinators do for us into monetary terms, although exactly what those terms are varies depending on the study. Research conducted at Cornell University found that in 2010 pollinators were responsible for adding $29 billion to U.S. farm revenues. Honey bees are the primary pollinators in commercial agriculture, and $19 billion of that $29 billion was thanks to them. Rightly or wrongly, over the latter half of the twentieth century, agriculture came to rely on honey bees. Farmers could bring in huge numbers of bees when needed and send them away when the crop was done flowering. They didn't have the worries and extra work of keeping bees themselves or providing out-of-season forage for them.

Since the 1950s, however, the number of managed honey bee hives in the United States has declined by fifty percent, while cropland needing bee pollination has doubled. The honey bees are being swept away by an avalanche of problems: hive beetles, wax moths, foulbrood, chalkbrood, stonebrood, *Nosema* fungus, Israeli acute paralysis virus, deformed wing virus (and about twenty other viruses), tracheal mites, *Varroa destructor* (another mite), colony collapse disorder (CCD), poor diet, pesticides on the flowers we ask them to pollinate, and maybe just plain old tiredness and overwork. What will we do if the honey bees can't keep doing the job? How important are they? Can the native bees step in and take over, or will we all be out in the yard pollinating our fruits and vegetables with our toothbrushes?

Bees and the sex lives of plants

After my "honey bees can't pollinate tomatoes" epiphany, I set out to learn more about pollinators. An estimated 200,000 different animal species pollinate plants, most of them insects, with bees leading the way. Now, bees are insects—there's no getting around it—and many of us don't like insects. Scientists have managed to identify and name around 900,000 species of insects, but they think that there are more unidentified species than identified ones. Estimates for how many unidentified species are out there ranges from 2 million to 30 million. (It's always good to give yourself some leeway, but we clearly need to be training more entomologists.) At any given moment, we likely share the world with about ten quintillion (10,000,000,000,000,000,000) individual insects. You could run screaming from this fact, but where would you escape to? They're everywhere.

However many species there are, every adult insect has three body parts cloaked in a hard exoskeleton: a head with two antennae, a middle section or thorax with six legs attached, and the rear end, which is called the abdomen. This last has always been confusing to me. In my mind, abdomens belong in the middle of a creature, but for insects the abdomen is the last part.

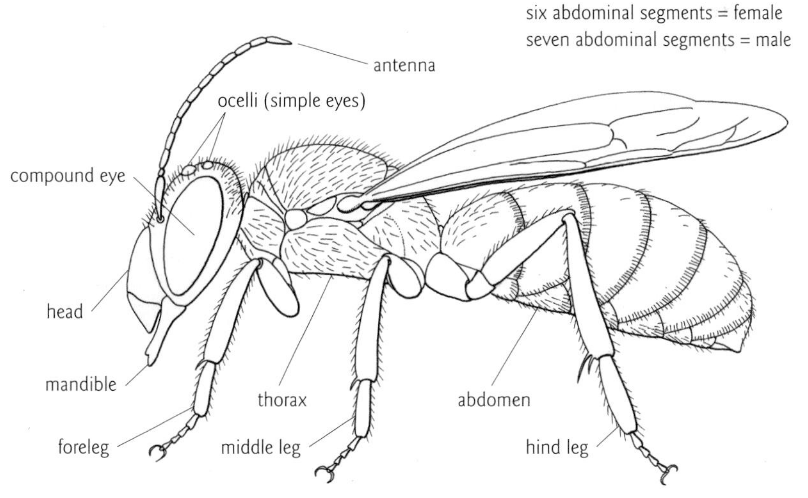

Parts of a bee.

Bees vary in size from the mighty (about an inch) like this carpenter bee, *Xylocopa virginica*, to the tiny *Holcopasites calliopsidis* that isn't much bigger than Roosevelt's nose on a dime. It isn't even the smallest bee in the United States. That award goes to *Perdita minima*.

The more I learned about pollinators, the more interested I became in the queens of pollination: the 20,000 species of bees worldwide that are largely responsible for the seeds of rebirth of three-quarters of the flowering plants in the world. I discovered that assuming, as most people do, that "bee" equals "stinging honey bee" was even more ludicrous than assuming "dog" equals "itty bitty Chihuahua." I goggled at close-up views of compound eyes that looked pixilated like a snake's skin; thick, luxuriant fur pelts; iridescent green and blue exoskeletons. I marveled at their diversity: bees with long tongues for slurping from flowers that hide the nectar deep, bees with short tongues that have no scruples about biting the side of those same flowers to steal nectar. I learned that the stinging potential of bees was vastly overrated. Males can't sting—they don't have the appropriate equipment—and females of most species don't bother to sting unless something

This female sweat bee, *Augochlora pura*, is from Tennessee. Sweat bees get their name because some lick up sweat. Those are her mouthparts sticking out the front of her head, and they clearly consist of much more than just a tongue. ▲

A female *Hoplitis fulgida* from Grand Teton National Park in Wyoming. This bee lines her nest cells with bits of chewed up leaves mixed with pebbles. ◄

Bombus sandersoni collecting nectar in Highlands, North Carolina. ▲

Bees have five eyes. Three are small and simple and two are compound, made of multiple facets that function as separate visual receptors. Only one of the three simple eyes is visible on the top of the head of this *Hylaeus modestus*, a masked bee, from Virginia. ◄

dire happens like getting caught in your clothes. It wasn't just the bees' appearance and diversity that fascinated me, but their facility at their jobs. Bees are literally built for pollination.

Bees diverged from wasps more than 100 million years ago and are simply wasps that went vegetarian. The bees' waspy ancestors (and their current wasp cousins) provided animals, living or dead, for their young to eat. Bees switched to feeding their babes plant pollen—easy to find and it doesn't fight back. Pollen is rich in proteins, amino acids, and fats, with some carbs thrown in as well. It's great baby food. Pollen also happens to be the flower's equivalent of sperm, setting up a fine opportunity for a mutually beneficial relationship.

Plants are immobile, which complicates their sex lives a bit. To mix up their DNA and avoid inbreeding, plants need to move their sperm equivalent, that pollen, around. For millions of years after plants made it out of the oceans and up onto the land, wind was the primary mover of pollen. Plenty of plants still do it that way. Pine trees, ubiquitous around my Georgia childhood home, rely on wind for pollen dispersal. The coatings on cars and swirls of yellow in the springtime gutters attest to the excess needed when relying on the serendipity of the wind to get your sperm to a receptive plant of the same species. That excess represents a lot of lost resources. Having something hand carry, if you will, the pollen is so much more efficient, but how to persuade someone to do the job? The usual way—payment.

Flowering plants provide that payment, and the method has proven hugely successful. Of the roughly 300,000 species of plants that grow on land, around ninety percent are flowering plants. Not all of these flowering plants have what we think of as flowers: buxom blooms like a magnolia or smaller ones like a dandelion. Grasses are flowering plants, and so are willows and maples, none of which have conspicuous flowers. What all flowering plants have in common is a seed (a baby plant in a nice weatherproof container) surrounded by—something. That something varies from a small dry coating not much bigger than the seeds themselves to a watermelon. Mosses, conifers of all sorts, and ferns are not angiosperms.

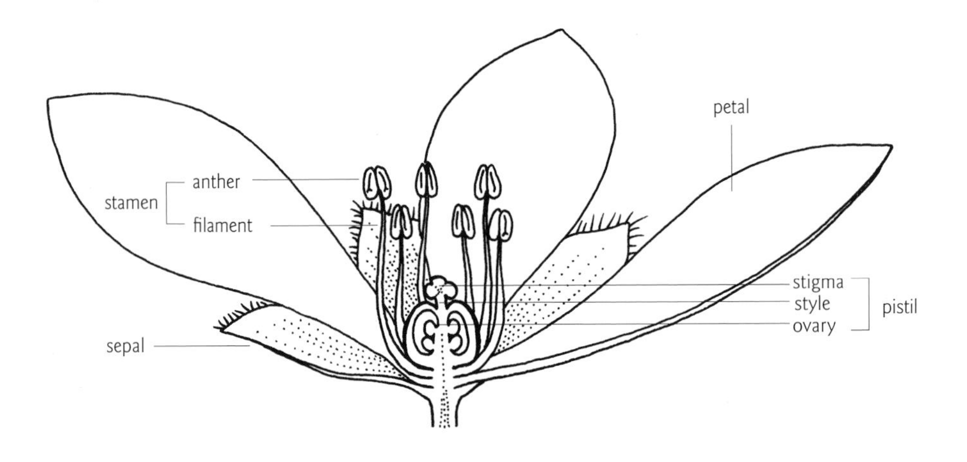

Parts of a flower.

All of our foods, barring a few oddballs like pine nuts and fiddleheads, come from the angiosperm clan.

Although angiosperms as a group evolved the payment plan to get animals to cart their pollen around, not every angiosperm works that way. Some are self-pollinated or pollinated by wind, but somewhere in the neighborhood of ninety percent of all angiosperms are thought to receive pollination services from animals ranging from lemurs, bats, and birds to moths, beetles, and flies. But it is the bees that have evolved to be the star pollinators. The payment the angiosperms supply these pollinators is nectar, a delicious, energy-laden carbohydrate drink. For bees they also provide those packets of baby food—pollen.

Most of the 20,000 species of bees worldwide collect pollen for their young to feed on, but about fifteen percent have evolved differently. Rather than collecting pollen, these mama bees sneak into other bees' nests and lay their young there, thus avoiding all the tedious work. Many of the bees that do collect pollen have special areas on their body for storing pollen and special devices on their legs for grooming pollen into those storage areas. Many bees have gotten hairier than their wasp ancestors, the better for pollen to glom onto them. When they fly, bees build up a slight positive electrostatic charge that sucks negatively charged pollen grains onto them when they land on flowers. Bees and pollen are made for each other.

Building a picture

When I was writing the proposal for this book, I had to explain why bees were so important and it made me a bit impatient. Bees are the primary pollinators of many of our plants. The world looks the way it does and we eat what we eat because of bees. What more needs to be said? I wanted to write a book that would seduce people with stories of bees, stories that would change their views of what a bee is and what they mean to us.

My hope is that after reading this book people will be so entranced by bees that they'll throw away their weed-n-feed and plant flowers in the lawn. I've traveled to farms, labs, and even golf courses hunting for the best bee stories. At the same time, I've worked to become a bit of a local bee nerd. I've spent hours helping on a local bee survey. I've taken bee identification classes. I've modified my garden, ordered bees online, and wandered around the city eyeing bees at work to see what plants Seattle's bees *really* like.

In addition to learning who some of our native bees are, I'd also hoped to answer some specific questions. Are managed honey bees truly in danger of expiring? If they do, can other bees step up and keep food on the table? How are the wild bees holding up out in the wild? It's all here—the education of one gardener about North America's bees. I hope the stories woo you as effectively as the bees have wooed me.

When I first came up with a title for this book I called it "Honey Bees Can't Pollinate Tomatoes: What Everyone Who Eats Should Know about America's Bees." Many of the stories in this book focus on bees in agriculture. What I've learned about how our food is grown has been as enlightening to this city girl as what I've learned about the bees.

I had barely started writing this book when people started thinking of me whenever they heard about bees. So when a truck full of fourteen million honey bees turned over on the highway near Seattle, I heard "I thought of you" from multiple people. And I smiled, pleased that they thought of me, but also rather wanting to wail, "Those are not my bees!" The bumble bees whose scientific name, *Bombus*, even sounds fat, the crazy alkali bees that climb into alfalfa flowers and get knocked upside the head again and again, the blue orchard bees that belly flop into flowers, the tiny enameled yellow and black *Perdita*—those are my bees. Yet I can't leave the honey bees out. They may have come from Europe, but they've been here long enough to get their naturalization papers, and they work awfully hard to keep us fed. Plus, this whole book began because of what they can't do. So before I delve into the world of our native bees, I need to talk about the honey bees, the gigantic migrant work force that supplies us with a good portion of our favorite foods.

A Bee for All Seasons

Apis mellifera, the European Honey Bee

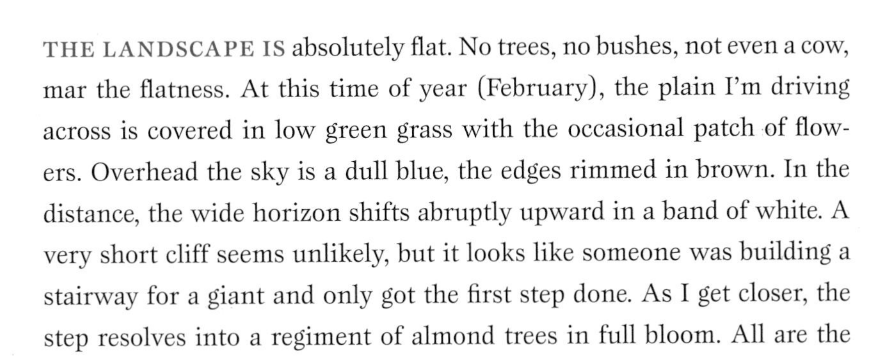

THE LANDSCAPE IS absolutely flat. No trees, no bushes, not even a cow, mar the flatness. At this time of year (February), the plain I'm driving across is covered in low green grass with the occasional patch of flowers. Overhead the sky is a dull blue, the edges rimmed in brown. In the distance, the wide horizon shifts abruptly upward in a band of white. A very short cliff seems unlikely, but it looks like someone was building a stairway for a giant and only got the first step done. As I get closer, the step resolves into a regiment of almond trees in full bloom. All are the same height, rising slim and straight from the plain, marching in perfectly spaced rows across the pale dirt of the Central Valley of California.

The almond army in California is a million acres strong, and this particular group is just one of the many I see as I crisscross the southern Central Valley from Lost Hills to Kettleman City to Fresno, Madera, and Los Banos. All but Fresno are just little towns or small cities hardly anyone has heard of surrounded by agriculture and empty valley grasslands. Not all the trees on this vast plain are almonds. Pistachios may march side by side with the almonds for a half a mile or more. I drive on through the flatness past a surprising patch of grapes, an empty tan field, and then more almonds—almonds of all ages. Some are just first-year sticks, but in only three to four years they too will be filled with flowers and starting to bear. The almonds I see are just a small portion of the acreage that runs 400 miles up the spine of California, producing more than eighty percent of the world's almonds and valued at $6.4 billion in 2013. And every dollar,

Almonds in bloom, and tumbleweed, in the southern part of California's Central Valley.

every bag of almonds or box of almond milk or chocolate-wrapped almond in a Hershey's bar is thanks to the biggest group of migrant farm workers in the world, the honey bees.

When I was researching this book, I ran across a quote by May Berenbaum from 2007 buried in some Congressional testimony. Berenbaum was Chair of the Committee on the Status of Pollinators in North America, and she told the members of Congress, "Even before CCD came to light, our committee estimated that, if honey bee numbers continue to decline at the rates documented from 1989 to 1996, managed honey bees will cease to exist by 2035." It floored me. Since then, I've learned a lot more about what honey bees face: disease, pestilence, famine, even flood and fire. Are we still marching toward the bee-pocalypse? I went to California for the almond bloom to try and find out.

Wonderful orchards

Gordon Wardell, the head of pollination services for the largest almond grower in the United States, stands at the end of a grove of almonds. Honey bees from twenty-four hives swirl around him as he takes one of the hives apart—with his bare hands. He wears a bee jacket with a hood but no gloves. I'm flabbergasted. He doesn't hop around, sucking on fingers beset by stingers. Instead, he pulls off a bit of honeycomb and asks, "Want to try some almond honey?"

"I don't really want . . ." my voice trails off as I hold up my gloved hands in explanation. I also wear a bee jacket, but I'm unwilling to take off those gloves or unzip my hood so close to the bees. Gordon kindly walks away from the billowing swarm so I can try this freshest of honey.

It drips. The comb is pliable, the honey thin and like no honey I've ever had; a bitter tang follows the sweet. Gordon asks, "Where else can you do that?" Nowhere in the United States, but I'm not sure if I'd want to again, considering the bitterness. Later, a commercial beekeeper will tell me that "almond honey is for the bees, not people." I have to agree, but that juxtaposition of good and bad seems appropriate given all the controversy surrounding almonds. The "nut" (which in botanical terms is a seed not a nut) is feted for its health benefits but reviled for its water use in drought-stressed California. And then, of course, there are the bees. Around 1.7 million hives, two-thirds or more of all the commercial bees in the country, come to California each February for almond blossom time. It's the most lucrative time of the year for commercial beekeepers.

Almonds may represent the biggest market for honey bee pollination in the United States, but plenty of other crops make use of honey bees: avocados and cherries, apples and blueberries, strawberries, melons, and more. How did we come to a place where so much of our agriculture relies on bees in boxes?

A short history of beekeeping

Humans love sweetness, and for most of history the only available sweetener in much of the world was honey. Few species of bees make serious amounts of honey because it's unnecessary. The adults of most bee species eat out at the floral cafeteria when they're hungry and only lay in stores (of pollen mixed with some nectar) for each kid. These bees have no need to have a larder filled for hard times because they usually die before the hard times hit, and their offspring wait out the hard times in diapause (a hibernation-like state) and don't eat.

Honey bees are different. During the summer they collect nectar and convert it to honey for long-term storage. A foraging bee sucks nectar into a special stomach, where enzymes start to break down the complex sugars into simpler ones. Then she comes back to the hive and passes the honey into the mouth of a honey-making worker bee. This worker swishes the nectar from mouth to honey sac repeatedly and then puts the now modified nectar in a wax cell, where she fans it to reduce the water content. The result is a food source that is nutritious and shelf stable.

My southern grandmother would say the bees are putting food by, only in tiny waxen cells rather than mason jars. This preserved food is for the queen and workers to eat during the winter because they don't enter diapause. Instead, they hunker down, shiver, and wait for the flowers to return. How much honey a hive will need for the winter depends in part on how cold and long the winters are. Forty pounds might be enough in mild areas and ninety or so where it's colder. These are quantities worth stealing.

And stealing honey is what humans have done for millennia. Cave paintings from 13,000 years ago show men climbing trees to steal honey. The Egyptians moved beyond thieving from bee hives in trees to keeping bees in containers, which is much more convenient for the people filching the honey. The Greeks, the Romans, and the Chinese all kept bees. The Mayans had no honey bees, but a native stingless bee (*Melipona*) made some honey, so they kept them. Not surprisingly, when the colonists came to North America, honey bees came with them to provide delicious sweetness and useful beeswax.

To put a honey bee colony in a container, you first have to catch it, and that is done when the bees swarm. The queen and workers wait out the first part of winter, shivering and eating their stored food, but once the days begin to lengthen, it's time to get back to work. The queen begins to lay eggs again, and the workers start caring for the resulting larvae. Numbers rise. As the flowers of spring come out and pollen and nectar are again available, bee numbers skyrocket. The hive fills, and eventually a tipping point is reached. It's time to swarm.

The queen lays some new queens and drones (there's no point in having these guys around until they're needed for sex because drones don't do any actual work). The old queen is the one who goes with the swarm, leaving one of her new daughters to take charge of the old nest. The old queen is too fat to fly, so the workers put her on a diet, feeding her less food. They start shaking and biting at her to make her move around the nest. Her egg production lessens, and surprisingly quickly—before the new queens emerge—the old queen is down to flying weight. The workers who are leaving also prepare, and they begin secreting wax for the making of the new comb.

When the moment is right, off they all go. The workers probably lumber, their bellies distended after they've fueled up on honey for their homeless period. The group flies to some nearby spot, often a tree branch, forming a beard of bees while they wait for scouts to search for a new hole to call home. It seems ludicrous, but while those bees are waiting they can be knocked off their perch and into some sort of box, and voilà, you've got a colony of bees. (Watch it on YouTube. It's pretty amazing.)

Bee swarms can alight in unexpected places.

When colonial-era beekeepers wanted to harvest honey or beeswax, they had to destroy (or at least significantly mangle) the hive to get it. Generally, the beekeepers would wait until fall and then smoke out the bees so they could safely get at the loot. Each year new swarms had to be captured. A few innovations were developed along the way, but the biggest step to modern-day beekeeping happened when Lorenzo Lorrain Langstroth invented the movable frame hive.

Langstroth was born in Philadelphia in 1810 and became a Congregational pastor and honey bee enthusiast. A photo of him in his eighties shows a round-cheeked old fellow with swept back white hair, Ben Franklin glasses, and a ministerial collar. He doesn't look like someone who spent much of his spare time mucking around with insects, but his book, *Langstroth on the Hive and the Honey-Bee: A Bee Keeper's Manual*, shows that is what he did. He is commonly referred to as the father of modern beekeeping because the hive he invented is still in use throughout the world.

Basically, a Langstroth hive is like a file drawer. The hanging folders are wooden frames filled with beeswax on which the bees build their comb. One of Langstroth's key findings was the distance needed between each of those hanging frames: about three-eighths of an inch. At that distance the bees can shimmy in and out between the frames. If the space is wider the bees build comb, and if narrower they fill it with propolis (a substance made from tree resin) and wax to seal things off. Each of these frames can be pulled out without destroying the hive. The bees can be checked and honey removed. A weak colony can be strengthened with additions from another or a strong one divided to make new colonies rather than letting the bees swarm.

All of this effort was aimed at getting honey and beeswax, not providing pollination, although the benefit of bees to pollination has been recognized for centuries. The Roman naturalist Pliny recommended bees for apple pollination. The first report that I could find of renting honey bees for pollination in the United States was in 1910. It wasn't until after internal combustion engines and good roads became common that migratory beekeeping, for honey or pollination purposes, really got going.

Melipona, the stingless honey bees

Bees have a range of lifestyles. Most in the United States are solitary; a single female works (and lives) alone to provide provisions for her babes. At the other end of the spectrum are the highly social bees like honey bees, which have perennial colonies with many bees and multiple generations living together. One queen lays eggs while the other bees have a variety of jobs, including care of the young. No other bees in the United States live like honey bees.

Just south of the border, however, is another group with a similar lifestyle. These are the bees in the genus *Melipona*, and they make honey too. Unlike the European honey bee, *Melipona* species are stingless. The ancient Mayans kept these bees for their honey, and their favorite species was *Melipona beecheii* because the bees were easy to manage and produced good honey (for a stingless bee). The European honey bee might produce ten gallons of honey, whereas *Melipona* would provide only around half a gallon. Nevertheless, the Mayans called the bees *xunan kab*, royal lady bees, and considered them a gift of the gods.

A *Melipona* female from Peru.

Nowadays, bees may travel from wintering grounds in Florida to the almond bloom in California to apples or cherries in the Pacific Northwest before heading to the Dakotas to load up on nectar for honey making. Some bees stay closer to home, but these days pretty much all commercial operations make a good chunk of their money from pollination. In 2016 the going rate was around $185 per hive for almonds (other crops will be less). It doesn't sound like much until you do the math. One Texas beekeeper sent 6000 hives to the almond bloom—that's a million plus dollars. The problem is what may happen to those hives while they are in the almond orchards.

Back in the almonds

Almonds are persnickety plants. They come from the lands of the eastern Mediterranean and require cold winters and hot summers, which are easy enough to come by, but they also need a very early, frost-free spring. In the United States, the only viable place for growing almonds is the Central Valley of California. Little blooms there in February, which means few native wild pollinators are available. Since the key to growing almonds successfully on a commercial basis is a whopping good pollination rate, California almonds are totally dependent on migratory honey bees for their success.

Gordon Wardell is a silver-haired, sixty-something fellow who hefts eighty-pound beehives with seemingly little effort. He's worked with honey bees throughout his career, and since 2009 he's been with the Wonderful Company. I show up at his lab on a sunny day in late February. To my surprise, the lab is not in a grove of blooming almonds but is instead set in the midst of a labyrinth of pistachio trees, another Wonderful Company product. When I turn off the main road to head to the lab, I'm greeted by a large sign warning me that this is private property and I was to neither trespass nor loiter. Loiter? I'd been driving thirty or forty minutes through a whole lot of nothing. Who'd drive all the way out here to loiter? I loiter only long enough to read the sign and then drive through the open gate and into a maze of trees. I make one turn, then another, and realize that getting lost is a real possibility. I'd seen buildings as I'd driven by on the main road, so I head in their general direction. I soon find a small empty building with a nearby warehouse and a giant one-story temporary building made of net—netting for the roof, netting for the walls—truly a net house.

Gordon is inside that net house and points to a door, a strangely house-like wooden door, set in the sea of netting. I walk through it into a tiny dark vestibule and grope for the door on the other side. I open this door—which leads into another vestibule. Clearly, no bees are getting out of this house. I open the third door and find Gordon waiting for me in Eden.

Five acres of plants spread before me, running into the distance in alternating stripes of green and brilliant blue—*Phacelia ciliata* in glorious bloom. I am dazzled by the beauty of the plants, the sun filtering down, the warm air, and I barely attend as Gordon explains the research project going on here. He's trying to find out what it takes to truly domesticate *Osmia lignaria*, the blue orchard bee (BOB). He talks of housing styles and locations, different nest tube diameters, various mud types for making the walls of nest cells. It's all interesting, but I really just want to go lie down amid all the wondrous blue flowers.

I understand why he and Wonderful are investigating *Osmia lignaria*. They are excellent pollinators of almond trees, and Wonderful probably wants to hedge its bets with the honey bees. In 2005 there weren't enough honey bee hives for all the almond growers who wanted them, and the U.S. government opened the international borders to a mass influx of bees for the first time since 1922. (The Honey Bee Act of 1922 had closed the U.S. doors to imported bees to keep out pests and diseases.) In 2010 the government mostly closed those doors again, so another shortage could cause serious problems for growers.

After we finish with the BOBs, Gordon and I head out to the almond groves about twenty miles away. We meet his wife, Terri, a teacher turned bee worker, and another bee worker, Natalie, to help remove pollen traps. When we arrive, Gordon, in a very gentlemanly fashion, holds up one of the white beekeeper jackets for me. I don't recall actually being asked how I feel about this, but I put on the jacket and zip it up. Terri gives me some gloves, and Natalie comes over and zips up a little gap I'd left undone, right by my throat, saying "This could be the difference between a good day and a really, really bad day."

I walk over to where Gordon is taking a hive apart, one of twenty-four set in a group. He takes off the lid and pulls out one of the wooden frames that dangles into the box and holds it up. Bees crawl around on the classic beehive hexagonal wax comb. Some of the hexagons are open, others have been capped.

At this time of year the colony is building brood (new bees) at a furious rate, primarily female workers. (Almost all honey bees are female workers, with only a few hundred males and a dozen or so queen eggs being laid each year.) The queen lays an egg in an empty hexagon, and three days later the egg hatches into a white grub-like creature, the larva. A larva that will become a worker gets fed royal jelly for three days and then is switched to the more plebeian bee bread (a fermented pollen-nectar mixture). Royal jelly is secreted from glands on a worker bee's head. This is clearly a specialty item because the difference between a larva becoming a worker or a queen is based only on how much royal jelly it's fed, with a queen-to-be getting royal jelly through her entire larval stage. A worker larva gets fed for six days total and then the cell is capped. It's time for the bee to go through metamorphosis. About twelve days after capping, an adult worker bee chews her way out, ready to get to work. So, it's three weeks from egg to adult worker bee.

Gordon points out various aspects of hive life. He shows me the queen, who is moving around pretty briskly. I'd expected something more slow and stately. I see a bee that is half in and half out of a cell. Gordon says, "She's nursing, she's feeding babies. They get about two hundred and fifty meals a day."

"Did you say two hundred and fifty meals a day?"

"Two hundred and fifty."

That's some pretty serious sisterly devotion.

Bees hum and whirl around us. Gordon slides the frame back into the box, still with his bare hands. I think about those bare hands as he works

Life in the comb: busy worker bees, white larvae curled in their cells, glistening nectar on the way to becoming honey, bee bread, and, at the top, some capped honey-filled cells.

with the bees, and later I ask him about getting stung. He says, "I do get stung all the time, it just doesn't bother me. I swell more from mosquito bites than honey bee. I'll get fifty to a hundred in a day. Those are bad days, and I'll eventually put on my gloves. But an average day, twenty to thirty stings. I got stung twice today. I just rub off the stinger and keep going. I didn't think to show you the stinger. It's fun to watch, though, because it pulses. The little muscle on top is putting the venom in."

Bee people are different from the rest of us.

Gordon pulls a pollen trap off the bottom of a hive box and shows me the day's catch: little yellow pellets of almond pollen, maybe half the size of a lentil. In a different part of the orchard, the pollen trap contains a pretty mélange of colors, showing that the pollen came from different plants. *Phacelia* pollen stands out, an exotic purple. The pollen is getting trapped for various research purposes, but I feel bad for the honey bees. They've been working their tails off, gathering pollen, grooming it tidily

Gordon Wardell taking apart hives at one of the Wonderful Company almond orchards.

onto the corbiculae (the bees' pollen baskets, flat spots on their hind legs surrounded by long hairs that are made to hold pollen), and wetting it down with saliva and nectar so it makes a tidy packet. The bees carry their booty home to the hive and crawl in through the bars of the pollen trap, where the pollen packets get knocked right off their legs and fall into the collector. All that hard work for nothing. Do they notice their loss? If not, imagine their confusion when they go to offload their packages.

Bees die

In a healthy colony, during the prime laying time the queen lays around a thousand eggs a day, mostly female workers. Twenty-one days later, a thousand adult bees chew their way through the wax plugs covering their cells, wait for their hair to dry and exoskeletons to harden, and get to work. Around six weeks later, all those bees are dead, replaced by new workers. It's how the honey bee world works until fall rolls around and the eggs of the so-called winter bees are laid. These bees need to be strong, not weakened by disease or poor nutrition or mites sucking out their blood.

A honey bee worker with a tidy packet of pollen on the corbicula of her hind leg.

Good winter bees are called "fat" bees and they need to live months not weeks. The winter bees are the ones that huddle through the winter and then go to work when the queen starts laying again as the days lengthen. They care for the brood and do the early foraging. A host of problems can throw off the normal death rate of a colony, and the untimely death of enough bees of either the summer or winter variety can wreak havoc.

When I was trying to fix a honey bee timeline in my head, I read old newsletters put out by Eric Mussen when he was the extension apiculturist at the University of California Davis. I started with 1994, the first year available online. The newsletters included all sorts of information, ranging from pesticide use to new honey bee products, but surprisingly often he wrote of bees dying and why.

- November–December 1996: Bees drown in heavy rain in California.
- May–June 1998: Significant losses of bees due to Captan (a fungicide) in almonds this year.
- January–February 2000: "We've been through this before, but perhaps not as badly. We've had 'autumn collapse,' 'disappearing disease,' 'spring dwindling,' and devastation caused by pesticides, tracheal and varroa mites. So, what caused the severe losses during the fall and winter of 1999?"

- March–April 2002: Disappointment and devastation in California as healthy fall colonies dwindle and die.
- March–April 2005: "Did we actually lose forty to sixty percent of the U.S. colonies this past season?"
- January–February 2007: Colony collapse disorder strikes.

Bees die. Clearly, they die from just about everything. They die from those two bee blood-suckers that arrived in the 1980s: tracheal and varroa mites. They die from viruses, some of which are passed on by the mites. They die from bacteria and fungi and hive beetles. They die from CCD (whatever the causes of that might be). They die from poor nutrition and pesticides applied to the fields they are pollinating. Hives are stolen and destroyed. Bees even drown every now and then. Honey bees die, and much of that death is an unavoidable part of doing business, but sometimes the deaths are maddeningly preventable.

Los Banos, California

It's early March, and the almond bloom is finishing. Gene Brandi, long-time beekeeper and president of the American Beekeeping Federation, has brought hives to his bee yard to wait until they head out to their next gig. It's raining and windy so the honey bees, being honey bees, stay home. Each morning when Gene checks on the hives, he finds the entrances clogged with dead bees. The worker bees on the death disposal crew aren't leaving the hive in this weather, so they just drag their dead to the exit and leave them. Over the course of the day, the wind and rain blow the dead bees away, but by the next morning there are more. If you opened the hive, Gene says, you'd see dead larvae, their tongues sticking out. They got confused and tried to emerge but only managed to chew off part of the caps covering their cells. The cause of death, according to Gene, is insect growth regulators (IGRs) applied to almond orchards. The IGRs bear no bee warning labels because when the product was approved the Environmental Protection Agency (EPA) required warning labels only

Honey bee sex

Apis mellifera, the European honey bee, is just one of the species in the genus *Apis*. Many other *Apis* species live in Asia, including some dwarf honey bees and the giant honey bee, *Apis dorsata*. All *Apis* bees living today make honey, set up permanent colonies, have morphologically distinct female castes (queens and workers), have hairy eyes, and the males have a huge and elaborate endophallus. Yep, we're talking bee penises.

The sex act for European honey bees goes something like this. The males hang out together, flying about above some prominent location like the top of a hill, waiting for their moment. A virgin queen flies by trailing her "come get me" scent, and the chase begins. She will usually mate with five or ten males while still flying high in the air. Each sex act lasts only a few seconds. The story of how that elaborate endophallus works is complicated, but the important part to know is that after the male ejaculates, the endophallus breaks, leaving a portion behind in the female. The male falls to the ground and soon dies from his few seconds of sexual bliss.

Apis florea, the dwarf honey bee, lives in the warmer parts of Asia ranging from Iran to Indonesia.

for products toxic to adult bees. Insect growth regulators, at least some of them, hurt the babes.

A few weeks earlier, during the middle of the almond season, I meet Gene for the first time when I hoist myself into his truck at my hotel in Los Banos. He's a middle-aged guy in a baseball cap, and he knows bees, beekeeping, almonds, water, alfalfa, cotton, pesticides. Gene grew up in Los Banos, and if it's related to honey bees in this part of the world, he knows about it. We drive out to one of the orchards where Gene has his bees, and it looks properly beautiful: white blossoms against a blue sky. The green grass between the trees is scattered with white petals. It's nothing like the dusty-looking orchards further south.

It's a good bee day—warm, not much wind, no rain—so the bees are out in force. As Gene and I stand in the orchard talking, we hear the sound of a small plane. Gene says, "That airplane is probably spraying an almond orchard."

I'm stunned. It's prime bee-flying time. I say, "One would hope they're not getting sprayed here while the bees are out."

Gene responds, "It happens."

I wonder if the sound of that plane causes a little surge of panic in Gene's stomach. Spraying during bloom in 2014 led to a massive bee die-off that made the mainstream news (including an NPR story entitled "Why are thousands of bees dying in California?"). Almonds, like other crops, have pests and diseases. One is a fungus that can get treated during bloom. Another is a pest that doesn't need to be treated during bloom. Nevertheless, sometimes an insecticide to control that pest, one of the IGRs, may get added to the tank with the fungicide. Why not? It's a free ride, after all. The IGR is meant to prevent pest insect larvae from becoming troublesome adults. Of course, bees are insects too, and studies have shown problems with the juvenile stages of bees after application of at least some of the IGRs that are out there.

You'd think the use of IGRs would be banned during bloom, like other insecticides that harm bees, but historically the labeling laws have applied only for harm done to adult bees. In 2014 adult bees weren't falling down

A honey bee flying in to visit almond flowers.

dead in the orchards; instead brood were dying in the hive. The reports in the news all stated that up to 80,000 hives had been harmed. Gene laughs at that number because it came out of an informal survey conducted in his brother's backyard. Gene knows that number didn't even include the losses for everyone who was in the backyard because the biggest beekeeper in the country turned in his form saying "still counting." Despite the heavy losses, no other count was ever done.

Gene says that growers have been using fungicides on almonds for as long as he has been in the business, but they only started adding the IGRs around 2000. He says no one knows what sort of synergistic effect there may be from mixing the two chemicals, but both fungicides and IGRs can affect colony health, even if they aren't harmful to adults. "Twenty-one days," he says, "egg to hatch for a worker. If that larva is fed IGR-contaminated food it may not develop right. It may not develop at all."

According to Gene, two "pretty high up EPA officials" attended the meeting in his brother's backyard. He says that when the EPA folks were asked to put some sort of warning on the labels, they replied that it would be years before these substances were up for review. The EPA has new guidelines on developing risk assessments for pollinators. Any new IGRs being evaluated now would be labeled as a risk to bees, but going back and reviewing all the old pesticides takes time.

The EPA folks said that rather than waiting for the review process, it would be better to come up with some best management practices, which the Almond Board of California did. These were available in time for the fall grower's meeting in 2014. Fewer bees died in the almond groves in 2015, so it looked like the growers had gotten the message.

Or perhaps not. Shortly after my visit in 2016, I emailed a few questions to Gene. In his response he wrote, "Brood damage has been severe this year for many beekeepers due to the bloom sprays of fungicides and IGRs. . . . I know of over ten thousand colonies negatively impacted at this point . . . not enough PCAs [pesticide control agents] and pesticide applicators adhering to the Almond Board's BMPs [best management practices]." (Gene conducted another voluntary survey later in 2016. Twenty-one thousand colonies were reported to be severely impacted by bloom sprays in the almond groves that year.)

Right before he sent that email, Gene went off to Washington, D.C., for a meeting whose goal was to increase communication between pesticide applicators and beekeepers. The focus of the meeting was how to get people to adhere to voluntary best management practices rather than depending on labeling that requires enforcement. All the profits an almond grower makes lie in the pollen packets on the hind legs of honey bees. If even almond growers can't adhere to best management practices, Gene says, "How will we ever get people to voluntarily accept anything?"

So, are the honey bees failing or not?

It's hard to know what to think about the status of honey bees, because so much of the information available is contradictory. Depending on when you pick up the newspaper, you may read CCD is ravaging bee colonies and the reason is neonicotinoid pesticides (neonics), cell phones (yes, cell phones), or Israeli acute paralysis virus. Then you read that it's not. You think all the deaths for the last ten years are due to CCD, but CCD has a very defined set of symptoms, and many of the deaths bees are currently suffering aren't a result of CCD. You read that the number of bee colonies is half of what it was after World War II and the number of beekeepers is dropping. The Obama White House saw the pollinator problem as one that "requires immediate attention to ensure the sustainability of our food production systems" and developed a plan and pledged millions of dollars to help pollinators. A few months after the White House plan came out, the *Washington Post* trumpeted, "Call off the bee-pocalypse: U.S. honeybee colonies hit a twenty year high." A few months *before* that article, a preliminary report stated that honey bee colony losses for 2014–2015 were over forty percent, with more than half of them coming during the summer, the time of plenty. You look at all this information and think, "What the hell?" So, what do we actually know?

For good or ill, managed honey bees are an integral part of how our current agricultural system functions, and they provide a good chunk of our favorite fruits and vegetables. Those bees, serving that agricultural function, are on life support from beekeepers.

I asked May Berenbaum about her statement before Congress in 2007 that managed honey bees would be gone by 2035 if they kept declining at the same rate seen toward the end of the twentieth century. She said that honey bees *aren't* declining at that rate anymore. "Beekeepers have stepped up their game," she said, and the federal government has responded as well. Berenbaum said there may be other catastrophes out there waiting, but at the moment things are holding.

So we have two potential problems: the health of the bees themselves and beekeepers' willingness to continue to keep bees.

Eric Mussen told me that just before CCD hit, the commercial beekeepers were hurting. The additional costs (extra feedings, treatments for mites, labor to split and re-queen, etcetera) weren't covered by the money coming in from honey and the low pollination rental rates (around $55 a hive at the time). But then came the bee shortage in 2004–2005, and the almond growers feared they wouldn't have enough pollinators to make a crop. Honey bees were brought in from Australia. Rental rates for pollination skyrocketed; growers were promising $120 per hive for the next year. Suddenly, the beekeepers had enough money to keep going. Right there is the key. Honey bees used in agriculture need beekeepers, and beekeepers have to be able to make enough money to stay in business. Every additional hit the bees take costs beekeepers more money, so pollination rates have to keep pace.

Honey bees are under assault and foundering from it. People work frantically to figure out what all the problems are and how to fix them. Some might blame the almonds. We shouldn't have so many of them. They use too much water. Much of the crop is sold overseas. If we didn't have almost all the bees in the United States showing up for the California almond bloom, maybe there'd be less disease and other problems spread around. If beekeepers weren't trying to get their hives ready to pollinate like gangbusters so early in the season when the bees would usually just be starting to build numbers, maybe life would be a little easier for the bees. Yet, varroa mites and some of the other bee problems wouldn't disappear even if the almonds did. And almonds are the biggest money-maker of the year for most beekeepers, allowing them to keep their bees going, do all the treatments, and have bees available for other crops and honey making. It's a dilemma.

So, we go on, looking for ways to fix things, both by helping the honey bees and by seeking other pollination options. Only a few other managed bees are available, and management issues exist for some of them as well. Or there's the 4000 or so species of wild bees out there. What role do they play and what role *can* they play in pollinating crops? A few more backups to our pollination system would be useful because nothing has really changed since 2012, when Jeff Pettis of the U.S. Department of Agriculture told the Steering Committee for the National Honey Bee Health Stakeholder Conference, "We are one poor weather event or high winter bee loss away from a pollination disaster."

Did Greenhouse Tomatoes Kill the Last Franklin's Bumble Bee?

AN OLD MAN walks the hills of southern Oregon, puffing on the slopes, eyes always scanning. He wears a T-shirt with a bee on it, a brimmed hat, and a photographer's vest stuffed with his tools: kill jars, collection jars, a camera, a voice recorder, and a kid's bug vacuum. He carries a bee net in his hand, ready for another day of tiny game hunting.

His prey is the Franklin's bumble bee (*Bombus franklini*), noted for having the smallest range of any bumble bee in North America and possibly the world. The old man is Robbin Thorp, the world's expert on Franklin's bumble bee. Robbin hasn't seen a Franklin's since 2006; neither has anyone else. The bee is probably extinct, but Robbin continues to hunt and to hope. On a warm August day in 2014, I join him for one of his hunts.

Robbin told me to meet him at the Mount Ashland Campground, up above the ski area. I arrive early, as usual, for fear of being late, so I wander. The campground appears to be built for mountain goats. Battered picnic tables stick improbably to the precipitous side of the mountain. The views from each site are spectacular, but I'd fear that a jaunt to the bathroom in the night would result in a quick trip to the valley below. I wade through dried grass and dying flowers, seeking bees. I see what may be a wasp and that's it. I'm not feeling hopeful about our chances this day. Soon enough, Robbin arrives and we set off, carting our gear.

Robbin Thorp hunting for Franklin's bumble bee on Mount Ashland in southern Oregon.

Bee hunting is a pleasant but rather random affair for something scientific. It consists mostly of moseying interspersed with brief periods of flailing with a net (beginners) or elegant sweeps (old pros). Robbin, a seasoned and wily bee hunter, has a graceful and economical bee-catching technique.

The sun is warm, the wind light. It's a good day for bees, assuming the hot dry summer has not put an early end to the bee season this year. We walk down the gravel road, shoes crunching, inspecting clumps of flowers for likely prey. Before long, the first catch of the day is in the net, buzzing angrily. To me, the bee is nothing but a black smudge obscured by netting. To Robbin, it is a male Van Dyke's bumble bee (*Bombus vandykei*). He reaches into the net and grabs the bee, pulling it out for inspection. Male bees have no stingers and so are safe to handle. I've tried to tell male from female bees before. It required a microscope and a dead bee.

It's the first of many bees we catch and Robbin identifies as we meander about the mountain. Robbin is a professor emeritus of entomology at the University of California Davis and has been studying bees for close to sixty years. Once I asked him how many bees he thought he'd identified over the years, and oh did he laugh as he replied, "thousands and thousands and thousands."

We are joined this day, as he often is, by other bee acolytes hoping to glean some of Robbin's knowledge, which he hands out readily. We all wander from flower to flower, a bit like the bees ourselves. Periodically I hear the whiff of a net swinging or the motor hum of Robbin's bug vacuum—tiny game hunting in action.

Robbin, ever the professor, quizzes us about the bees we find, but as the day progresses, conversations become more terse.

"I caught somebody big," I say.

Robbin walks over and contemplates the bee briefly. "That's a queen of

The Franklin's bumble bee, possibly the first U.S. bumble bee to go extinct, is now only seen pinned in museums.

flavifrons," he says. "Where'd you find that? You can see a couple of yellow bands and the black on the tip." We release her and Robbin records the find on his voice recorder. We wander some more.

We head down to the Pacific Crest Trail and amble along it for a while. Somewhere along this stretch is where Robbin saw the last known Franklin's bumble bee eight years earlier. I forgot to ask where, my brain mush after hours of trying to recall bee markings. I'd wanted to see the exact spot, but maybe it doesn't matter. What does matter is that the Franklin's bumble bee seems to be gone, likely for good.

On bumble bees

Most people's knowledge of bees revolves around honey bees—their intricate social lifestyle, their honey making—but that's not how most bees live. Most bees live their short adult lives (a month to six weeks) alone. Males mate and soon die. Females mate, provision cells for eggs, and die. Some species *are* social, with a band of related females living and working together, but none native to the United States and Canada are as organized and impressive as the honey bees. Those in the genus *Bombus*, the fat and charismatic bumble bees, are among the crew that is social. Bumble bees differ from honey bees in several ways. They don't have a sophisticated means of communication like the honey bees who do a little dance to show the way to good pollen and nectar resources. A bumble bee queen can survive without her workers; a honey bee queen can't. Also, a bumble bee colony lasts only a season, not the years that a honey bee colony can survive.

In North America, around forty species of bumble bees form family groups that last one growing season. Another six bumble bee species don't form their own colonies. These are the cuckoo bees, invading the nests of other bumble bees and forcing the workers to rear their young. Only the young mated females who will start the following year's nests (or, for the cuckoo bees, parasitize the nests) make it through the winter. This differs from a honey bee hive, where a queen and a core of workers survive the winter on food stored away for that purpose.

Much less effort has been expended on understanding bumble bees than honey bees. The grandfather of bumble bee research is Frederick William Lambert Sladen, born in 1876 in Dover, England. As the son of an affluent family, he was tutored at the family home, Ripple Court, and clearly had few chores that interfered with his bee observations. In 1892 he wrote a forty-page pamphlet on "humble-bees," as they were then known (because they hum when they fly), that pretty much established him as the world's expert. He was sixteen. Sladen continued his bee research, and in 1912 he published *The Humble-Bee*, 278 pages of evocative, turn-of-the-century prose.

"The story of the humble-bee," writes Sladen in *The Humble-Bee*, "is largely that of the queen. From start to finish she is the central and dominating personage upon whose genius and energy the existence of the race depends." He notes, "At first her duties include those of the workers, her brood depending upon her for everything—food, warmth, and protection from enemies. She nurses it with as much motherly devotion, industry and patience as we see displayed by many birds and mammals in the care of their young." He includes details upon details: the sweet smell of male bees (which he caught and sniffed) and theories on why they smell so, and from whence the smell comes (their mouths). He explains, with drawings as well as words, the special devices on a bumble bee's legs for cleaning antennae. He writes at length on the care and feeding of the young at all stages of their lives. He tells of the end of life and how the queen loses her hair as she ages, how old queens of certain species die a peaceful death, sitting with the last of their children on top of the waxy remains of their lives. The book is a marvel of love and information.

Based on Sladen's research and that of a few others, a generalized bumble bee season goes something like this. A queen-to-be, impregnated the previous fall, wakes in spring or early summer (depending on the species), famished after living off her own body during her winter hibernation. She hunts for food and a good nest spot—a hole big enough to hold dozens to hundreds of bees—often an abandoned rodent hole. Once she finds a good spot, she starts laying in stores of pollen and nectar, the latter going into little pots that she makes from wax secreted from glands on her abdomen.

She lays the eggs on the pollen, covers them in wax, and then spreads herself on top to keep the eggs warm. The eggs hatch in about four days as white, legless, motionless eating machines. For about ten days they eat and grow, with mom leaving the nest only to gather more food for their ravenous mouths. Then the larvae spin cocoons and undergo that magical transformation known as metamorphosis, which for a bumble bee means turning from a white maggoty thing to a flighted fuzzy bee.

The first hatchlings are female workers (by definition a worker bee is a female bee). When they emerge, the queen no doubt heaves a great sigh and says "get busy." The new bees help with chores around the nest and go out for the groceries, while the queen settles into egg laying. And so the season goes. Eggs that will become males and the following year's queens are laid toward the end of the season. Some of the workers may lay male eggs at this time. In at least some nests, civil strife breaks out toward the end of the season. As Sladen puts it, "In a [*Bombus*] *lapidarius* nest a strange scene may be witnessed at the laying of the male and queen eggs. The workers, hitherto so amiable, are suddenly seized with anger and jealousy, for as soon as the queen has closed the cell and turned away, one or two of them hurriedly commence to bite it open, their wings quivering with excitement." The queen will have none of this and rushes back, beating them away. Clearly, bumble bees haven't perfected living together as well as honey bees have.

A lot can go wrong before the next generation of queens is hatched. A queen-to-be may have chosen a bad spot to hibernate and get washed out by winter rains. A long stretch of bad weather after the bee wakens that prevents her from going out may result in her starvation. A sudden change in the weather on a foraging trip can leave a bee stranded and exposed, too cold to fly home. A dearth of good holes for nesting can lead to queen battles for the rights to the hole, and the death of the loser. A young nest can be eaten by ants or mice. Cuckoo bumble bees can enter the nest, kill the queen, and force the workers to look after their own offspring. The young may be eaten by the caterpillars of wax moths. Insecticides may kill bees, and herbicides may destroy their food plants. It's an uncertain world for a bumble bee.

Sladen did his work on British species of bumble bees. No one has done the same for the Franklin's bumble bee. We don't know about its specific housing and lifestyle needs. Even Robbin isn't certain where Franklin's likes to nest. At least we know what it looks like. The bee Robbin seeks is a typical bumble bee—big, fat, and fuzzy—but even as bumble bees go, Franklin's is one of the bigger ones. Female workers are up to half an inch long, and the queens push an inch. Most bumble bees are showy things with snazzy stripes and patches to advertise their stings, but Franklin's is a bit on the dowdy side. Its back end is a sober black, although it does have a particularly big yellow patch on the back of its middle section, reaching down between the wing bases.

Even after studying diagrams and pictures, bumble bee identification in the field for novices falls somewhere between head-scratching uncertainty and absolutely maddening. Bee identification cards and books show little idealized colored drawings of what the bees look like. Each type of bee—queen, worker, and drone—gets a line of their own for each species. Most regions will have at least a handful of potential bumble bees to choose from, so you figure out who lives in your area and get the charts lined up. And that's when the dismay is likely to set in. Males often don't look like workers, and both may be different from queens. Even worse, the charts show several different appearances for each type of bee. For example, the chart for the yellow-headed bumble bee (*Bombus flavifrons*) shows six versions of the queen, six of the worker, and seven of the male. Multiply that by the handful of bumble bee species in the area, and you feel like just giving up.

Fortunately, as I eventually discovered, a lot of that variation turns out to be irrelevant because the differences are usually regional. A yellow-headed bumble bee from New Mexico may not look like one from Seattle, but the New Mexico bees in each group (queens, workers, and drones) will probably all resemble each other pretty closely. Franklin's bumble bees are easy, with only one version of the queen, two of the workers, and one of the male.

So, for most of us, bumble bee identification goes something like this:

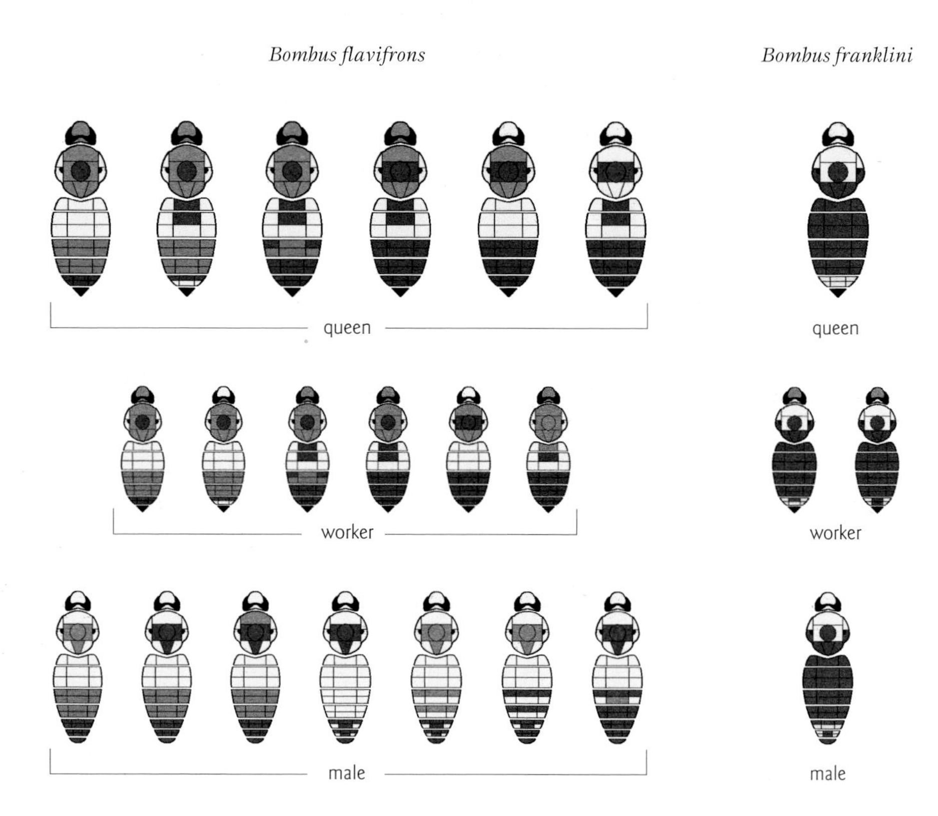

Some bees have a lot more variation in their appearance than others, but much of the variation is geographic. *Bombus flavifrons*, with a range encompassing a good portion of western North America, shows much more variation than *Bombus franklini*.

catch the bee, put it in a collection jar, pop the jar into a cooler to slow the maddened bee down, and then haul it out and try to compare the relevant markings to what we expect from our local bees. Robbin, after all these years, can just tell. "I was out in the field with Virginia Morrell, who does writing for *National Geographic*," he said. "We were standing on the road, talking about Franklin's—this was back when it was still around—and about fifty yards down slope I spotted a bee and said, 'That's Franklin's!' and ran down the slope and picked it up and sure enough it was."

Historically, Franklin's bumble bee wasn't particularly special, except for its small range. It was a fat, fuzzy bumble bee just doing its thing. What has made Franklin's bumble bee special is that in less than ten years it seems to have gone extinct.

In the beginning

The Franklin's bumble bee story begins, unexpectedly, with the northern spotted owl. In 1990, when the northern spotted owl was listed as a threatened species under the Endangered Species Act, huge tracts of forest in Oregon and Washington were closed to logging. Uproar resulted, lasting for years. Within the range of the northern spotted owl lies that of Franklin's bumble bee, another potential candidate for the Endangered Species List due to the risk of habitat destruction in its tiny range. Every Franklin's bumble bee has historically lived within a 200 by 70 mile oval, elongated north-south, centered around Ashland, Oregon, near the California border. The U.S. Forest Service controls a lot of the land in that part of the country. In 1998, the agency contacted Robbin to perform a survey of Franklin's bumble bee, thinking, according to Robbin, that another endangered species in the area would take some of the heat off the northern spotted owl.

So for the past eighteen summers, Robbin has tramped the hills of southern Oregon and northern California hunting for Franklin's bumble bee.

When Robbin began his Franklin's bumble bee survey in 1998, all appeared well. He found ninety-four bees, which seemed like a reasonable number to him, middle of the pack for bumble bees in that area. The next year he found only twenty, but thought nothing of it; bee populations vacillate. The next year, 2000, he found nine, the following year one. A blip of twenty in 2002 and then a dive back to single digits. Trouble.

Franklin's bumble bee wasn't the only bee in trouble at the time. Three other species, the western bumble bee (*Bombus occidentalis*), the yellow-banded bumble bee (*Bombus terricola*), and the rusty-patched bumble bee (*Bombus affinis*), all of which are in the same subgenus as Franklin's

Bumble bee lookalikes: Is it even a bee?

The insect in the photo screams bumble bee—fat body, lots of hair, stripes—but I can't figure out which bee it is. Eventually, I admit defeat, and send the picture off to one of the instructors at the bee class I'd taken. Her response is kind, "[It] is a fly, and a nice fuzzy one." Well, rats.

Fortunately, once you start paying attention, it's pretty easy to sort bees from flies. Bees have four wings, flies have two. That distinction isn't always as helpful as one would think because if the wings are beating fast, you can't even see them. Also, bees can hook their fore and aft wings together so they look like one. Another clue is the eyes. Bees have biggish eyes off to the sides of their heads, and flies have great big ones right on top. Bees have long antennae, and flies have little short ones. Female bumble bees have stout hind legs with a flat spot for storing pollen, and flies have scrawny legs.

So, once you've sorted the flies out, you're only left with all the other bees that look like bumble bees.

This syrphid fly, *Volucella bombylans*, in El Cerrito, California, does a great job as a bumble bee mimic. I call them wannabees.

bumble bee, experienced precipitous declines in the late 1990s and early 2000s. These other bees had much larger ranges than Franklin's bumble bee, and only the range of the western bumble bee even overlapped with Franklin's. Clearly the problem wasn't one of place: pesticides or habitat issues. Whatever the problem was, it was hitting only closely related bees, but doing so all across the country. This is where the tomatoes come in.

Bumble bees and tomatoes

Bumble bees are the preeminent pollinators of tomatoes, because they, unlike honey bees, are capable of buzz pollination. Now, pollination is just plant sex, moving the sperm equivalent (pollen) to the female parts (the flower's stigma). On most flowers the pollen is out at the end of a stalk coated onto little paddles or balls called anthers, ready for the taking. However, some flowers, including those of tomatoes, hide their pollen *inside* the anthers so it has to be shaken out. A bumble bee does this by grabbing the flower in its mouth, curling its body around the anthers and rapidly contracting muscles in its thorax, causing vibrations that shake the pollen right out of the tiny holes in the anthers, like shaking salt from a shaker. This activity makes a buzzing noise, and so, buzz pollination.

Growing tomatoes out of season in greenhouses has long been a lucrative business. Records from the 1890s show wintry tomato prices at forty to eighty cents per pound, which would come to around $10 to $20 per pound in today's dollars. The problem is that there is no wind or natural pollinators in a greenhouse to shake out the pollen, so people have resorted to all sorts of methods to find the best way to pollinate greenhouse tomatoes. In "Experiments in the Forcing of Tomatoes" (1891), Liberty Hyde Bailey writes, "The common practice is to tap the plants sharply several times during the middle of the day with a padded stick. This practice is perhaps better than nothing." He did offer other, better, options. One could try hand-pollination with the spoon or slide method: shaking pollen into a little spoon or onto a slide and then dipping a flower into the pollen. Sometimes that didn't work so well, and it was thought better to emasculate the

Two bumble bee lookalikes, *Xylocopa* and *Anthophora*

To tell a bumble bee female from its lookalikes, look at the legs. (This only works for female bees, because the males don't carry pollen back to the nest.) Most types of bees stash that pollen on their hindmost legs for transport, but they don't all carry it the same way. Bumble bees have a flat shiny spot on the outside of their leg surrounded by long hairs. They wet pollen down with saliva and pack into tidy wads. With or without the pollen, the legs are distinctive if you can get a good look at them. Both *Xylocopa* and *Anthophora* have hairy hind legs with no flat spots. They carry pollen, dry, in the hairs. *Xylocopa* also differs from the other two genera by having a shiny rather than a hairy abdomen. It's a place to start.

Xylocopa virginica, the eastern carpenter bee, looks a lot like a bumble bee. Look to the legs and abdomen to see the differences.

Although this *Anthophora californica* female resting in El Cerrito, California, looks a lot like a fat fuzzy bumble bee, her pollen-carrying structures show otherwise.

flower—remove the petals and anthers (yep, ripping the male parts right off)—so the stigma was obvious and one could be sure of getting the pollen on the stigma *and* be sure of not wasting time trying to pollinate the same flower again. Later on, people tried shake tables, blowers, and vibrators, all this work to get those high-value, out-of-season tomatoes.

While some people spent the bulk of the twentieth century figuring out the best way to grow tomatoes in greenhouses, other people were trying to figure out how to domesticate bumble bees. Now, a domesticated bee is one for which people control some aspects of the bees' lives, usually with the ultimate desire of controlling when and what they pollinate. It's not as simple as it sounds.

To domesticate bumble bees, people had to figure out how to wake a hibernating queen bee and get her to start making workers on demand and out of season. It was the work of decades. What temperatures should hibernating queens be kept at? What temperatures and light levels should be used to break hibernation? How to get her to actually start laying eggs? (They tried putting two queens together, which certainly got everyone awake and ready to go but usually ended up with at least one dead queen.) How do you get queens to mate? People got nests going in various ways, but it was tricky and labor intensive and mostly done on a small scale for research purposes. Nobody seemed to have grand commercial plans for waking up hordes of bumble bees out of season.

Then, in 1985, Roland de Jonghe, a Belgian vet and bee enthusiast, put all the bee-rearing information together with the fact that a colony of bumble bees could pollinate a greenhouse full of tomatoes way better than a bunch of people with vibrators or blowers. Since bumble bee nests only last a few months, people would need new nests every year, and de Jonghe saw the business potential of those bumble bees. In 1987 he started the first business rearing bumble bees for commercial use. He used a European bee, the buff-tailed bumble bee (*Bombus terrestris*). A few other European companies joined in, and by 2004 close to a million colonies of bumble bees pollinated nearly 100,000 acres of greenhouse tomatoes worldwide, with an estimated value so large that I thought it was a typo—except it was written the same way twice—12,000 million euros (about $14 billion) per year.

Greenhouse tomato growers were ecstatic about the bumble bees, no more costly and time-consuming hand-pollination. U.S. growers wanted in on the bumble bee revolution, but a section of the U.S. Department of Agriculture, the Animal and Plant Health Inspection Service (APHIS), said no to early applications for importation of the European bees. (APHIS's job is to protect agriculture, and keeping out plants, animals, and diseases that might cause a problem is part of its mandate.) In the early 1990s two native bumble bees, *Bombus impatiens* and *Bombus occidentalis*, were being reared in North America, but one of the companies involved wanted to raise the queens in their European facilities instead, sending the resulting young colonies back to the United States to get to work. APHIS agreed. From 1992 to 1994 these European-raised North American bees flew home across the Atlantic with their offspring, ready to pollinate. *Bombus occidentalis* went to its home range in the west, and *Bombus impatiens* to its range in the east. Returning with these bees, hypothesizes Robbin, was a European disease that escaped into the wild. All the bee species that declined are from the same subgenus as *Bombus terrestris*, the European bee being raised for commercial purposes in Europe—including Franklin's bumble bee.

Cause and effect?

The first commercial North American bumble bees came back from their European treatment in 1992. Within fifteen years, four species of bumble bees, over eight percent of the bumble bee diversity in the United States and Canada, had been decimated. Is it cause and effect or coincidence? Although no Franklin's bumble bees are available on which to test hypotheses, studies show that Robbin's imported disease idea is plausible. Commercially raised bumble bees tend to have higher levels of disease than wild bees, and research indicates that those diseases have escaped into wild populations near where the bees have been used. A 2016 study into a likely culprit, the fungal parasite *Nosema bombi*, found no definitive answers. The Franklin's bumble bee's homogeneous appearance suggests minimal genetic diversity, which might explain why they seem to have

A bumble bee visiting a greenhouse tomato. ▲

Bombus impatiens, the eastern bumble bee, is the only bumble bee currently in commercial production in the United States. ◄

Other bumble bees in decline

The abundance and range of four other species of North American bumble bee, all widespread and common, also appeared to be shrinking over the last twenty to thirty years. Data were insufficient to know for sure, so a group of researchers decided to find out if the bees that seemed to be declining were indeed declining.

Between 2007 and 2010, the group went out and collected eight species of bumble bees, four thought to be in trouble, four not. Three of the four thought to be in trouble are close relatives of *Bombus franklini* (*Bombus terricola*, *Bombus occidentalis*, and *Bombus affinis*), while the fourth (*Bombus pensylvanicus*) belongs to a different subgenus. At the end of the three years, the team had collected 16,788 bumble bees, but to know if the bees were in decline, they needed some numbers to compare them to. They went to natural history museums throughout the United States and developed a database of more than 73,000 bees. The four species thought to be in trouble were indeed declining, both in abundance and distribution, with the declines occurring in the last twenty-five to thirty years.

Bombus occidentalis, the western bumble bee, is another close relative of Franklin's and has mostly disappeared from the westernmost part of its range, with a total estimated range reduction of twenty-eight percent. ◄

Bombus pensylvanicus, the American bumble bee, is in a different subgenus than Franklin's bumble bee, but it too has seen a range loss of about twenty-three percent. ▲

Bombus affinis, the rusty-patched bumble bee, has seen the worst reduction, other than that of Franklin's itself, with a range loss estimated at eighty-seven percent. ▲

Bombus terricola, the yellow-banded bumble bee, is a close relative of Franklin's bumble bee and has seen a range reduction of thirty-one percent. ◄

been completely wiped out while the more widespread and diverse bees still show some flickers of life.

The western bumble bee only lasted for a few years as a managed bee before succumbing to disease. The eastern bumble bee thrives and today is the only commercially available bumble bee in the United States. When this all began in the 1990s, everyone involved had worked out a gentleman's agreement that movement of bees outside their native ranges, even within the confines of the United States, was a bad idea. In 1994 Congressman Sam Farr of California was concerned that this agreement might change, and he wrote to then Secretary of Agriculture Mike Espy about it. Espy replied that both *Bombus occidentalis* and *Bombus impatiens* were being kept in their native ranges and that APHIS had conducted risk assessments that indicated that allowing *Bombus impatiens* into the west might bring along eastern pests and diseases. But then the commercially raised western bumble bees died out and growers in the west didn't want to go back to the bad old days of hand-pollinating tomatoes. So they started agitating for the eastern bumble bees. In a briefing paper from 2009, APHIS reported that the position change was "due to pressure from greenhouse growers," leaving any restrictions up to the state involved.

In 2010, with plenty of data in hand, the Xerces Society for Invertebrate Conservation, along with the National Resources Defense Council, the Defenders of Wildlife, and Robbin Thorp submitted a petition to the U.S. Department of Agriculture requesting controls be placed on the interstate movement of bees. And then they waited. In 2013, the group pressed for a response. According to Robbin, the U.S. Department of Agriculture replied that they had not shelved the idea, that they would take an evidence-based approach, and were seeking funding to study the problem. No timeline for a decision was given. I contacted APHIS in 2016 and was told that they require evidence that a clear benefit would result from stopping the interstate commerce. Since there currently aren't sufficient data in the literature to make that decision, APHIS has commissioned some studies. Now, a new president lives in the White House, and with him comes a new set of priorities that may not include funding for studies on the safety of moving bees outside their native ranges. So, we continue to wait.

While tomato growers campaigned for bees to be moved outside their native range and various researchers investigated what had happened to those other bees in decline, Robbin continued his dogged advocacy for Franklin's bumble bee. On June 23, 2010, Robbin and the Xerces Society submitted a petition to list Franklin's bumble bee as an endangered species. When I checked the Endangered Species website in April 2015, I saw the petition listed as "under review." The review should have been completed by September 2011, but the portion of the U.S. Fish and Wildlife Service that deals with petitions to the Endangered Species Act was drowning in both petitions and litigation at that time. Things were so bad that, in 2010, agency officials said they had spent so much of their $21 million budget on litigation and responding to petitions that they had little money left to protect species. In 2011, a settlement was reached with the two environmental groups who were responsible for the bulk of the petitions and litigation. U.S. Fish and Wildlife submitted a work plan to the U.S. District Court saying they would clear up the status of 251 backlogged species within six years. The Franklin's bumble bee was not part of that group, and, once again, we wait.

The last Franklin's bumble bee

Robbin Thorp is in his eighties now, but several times a summer he gets in his white truck and drives the four to five hours north into the small oval of land where Franklin's bumble bee once flourished. There, he walks the hills looking for that black-bottomed bee. I asked him why he still goes out looking for a bee that's likely extinct and that hardly anyone else cares about. He has many reasons: meeting again with the bee-hunting contacts he's made over the years, the views he gets of Mount Shasta from some of his field sites, escaping the summer heat of Davis. But "the main reason I keep going," he wrote to me, "is the diminishing hope that I may again see this magnificent critter and watch its recovery unfold as I did its dwindle."

Robbin saw the last known Franklin's bumble bee on August 9, 2006, just off the Pacific Crest Trail on Mount Ashland. "I was wandering around," Robbin says, "intently searching for Franklin's and occidentalis [the western bumblebee] and just happened to spot one [a Franklin's] on sulfur eriogonum. It's not one of the more common hosts, but I had seen it on that plant before. It paused a little bit, and before I could get my wits about me, I didn't have my camera or my little ice container to chill it down, it began to fly off, so I tracked it as far as I could. I kept revisiting that site during the rest of the day to hopefully see her come back, but never saw another sign of her."

Osmia lignaria, the Great and Glorious BOB

THE LITTLE BOX sits on the front table, standing on end as though waiting to greet me. I look at it in confusion and then remember—my bees. Oh crap, it's my bees, my hibernating blue orchard bees (BOBs), which have been sitting in the warm house all weekend while I was out of town. I put the box to my ear: no angry buzzing. Relieved, I pop the box in the fridge to deal with later.

Once I have some time, I fetch my bees and head out into the cold January night because I'd had this inner vision: I slice the tape and a swarm of angry bees flies out like a scene from a horror movie. Even I, lover of bees that I've become, still have thoughts like this pop into my head. I know it's stupid. For one thing, these bees need to be at least 54°F (12.2°C) to fly, and they were just in the fridge. For another, these are BOBs (or orchard mason bees, *Osmia lignaria*) and BOBs don't do angry; it's not in their nature. Nevertheless, outside I go.

I open the box and—nothing. I peer inside and see an even tinier box, only a bit bigger than a match box. I pry it open and pour out twenty cocoons the color of pencil lead, placing a few on my palm. I look at them, astonished. They have no perceptible weight, yet inside each sleeps a fully grown BOB. I know female BOBs are about the size of a honey bee and are larger than the males. Hmmm. These little lozenge-shaped cocoons, the biggest maybe a bit larger than a Tylenol capsule, don't look big enough to hold a bee the size of a honey bee. Maybe my bees are runts? Maybe last year was a bad year for bees and the moms weren't able to lay down adequate food stores for the babes to grow big and fat? Or maybe the bees are just stuffed in there really tight? A few cocoons are way smaller than the others, more like the size of two rice grains. Could the males really be that much smaller than the females? All my book knowledge seems useless as I look at these cocoons and wonder at their inhabitants. At least all appears well with them: I see no tiny holes suggesting the bees are awake and trying to get out.

It's January, too cold here in Seattle for even the hardy BOBs to be flying, so I put the cocoons back in their tiny box, end open. The tiny box goes back in the bigger box along with a cotton ball soaked in sugar water. I'd read somewhere about adding the cotton ball as a potential food supply, although it doesn't make sense because the cold of the fridge should keep the bees dormant. I figure the cotton ball can at least act as a humidifier. I warn my children there are bees in the refrigerator. My son accuses me of trafficking in bees, while my daughter just refuses to open the fridge. They do not share my fascination with bees.

Dave Hunter and the great BOB experiment

Although the bees and I must wait a bit longer for warmer weather and the first flowers, Dave Hunter of Crown Bees in Woodinville, Washington, isn't waiting. Dave is in love with BOBs. He has a plan, a truly monumental plan, to use BOBs to help the beleaguered honey bees and our food system by fostering a partnership between gardeners, farmers, and BOBs, with Dave acting as the bee-broker middleman.

I first meet Dave at his office in suburban Seattle. I expect a bee factory to be a dusty, corrugated steel warehouse, but instead I find Dave in a small clean office in a tidy strip mall next to places like Crossroad Signs and Builder's Interior. Dave thinks BOBs are bees with little downside. They're top-notch pollinators, they almost never sting, and, compared to honey bees, they are easy for novices to raise. Dave was in the engineering world before he started this bee business—he also sells leafcutting bees,

Osmia lignaria goes by many names: orchard mason bee, blue orchard bee, and BOB.

bumble bees, and bee accoutrements—and he brings a sense of engineering practicality to the job. When some of his bees were dying in the cocoon from desiccation, he developed the Humidibee in which to store them. The honey bees are in trouble? Dave comes up with an enormous, yet simple, plan to help them.

After we talk for a while, Dave takes me into the back room, an appropriately warehousey space full of sleeping bees and sundry bee paraphernalia. This, Dave informs me, is the headquarters of the biggest BOB company in the world. I look around, able to see into every corner from where I stand. Clearly, any grand ambitions for having BOBs help save the honey bees are still in their infant stage.

For 2015 Dave thinks that he and the other purveyors of BOBs have about a million bees to sell, up from around 800,000 in 2014. When I ask Dave how many BOBs he thinks will be needed to help the honey bees, he replies, "About a billion."

The Life of BOB

I stand in an orchard of cherry trees, looking for BOBs. One lands nearby, climbing into the hole of a wooden nest block. The bee is about the size of a honey bee but looks nothing like one. It is shiny and black, and its three body parts are very round. One bee person I know calls them BB bees since they look like three good-sized BBs all lined up. To me, they are the muscular mesomorphs to the bumble bees' plump endomorphs and the honey bees' svelte ectomorphs. Also, the black seems to lend BOBs a stealthiness that makes them easy to overlook. I know thousands of BOBs are flying in this orchard, but I see few. Are they off to the hinterlands looking for tasty weeds? Perhaps they are swooping around, hiding in plain sight—black stealth pollinators.

It's not just their appearance that differentiates BOBs from honey bees, it's their behavior as well. Honey bees are the bankers of the bee world, working short hours and taking all the holidays off. If it's raining, they go home. Too cold? They don't even leave the hive. BOBs, on the other hand, start flying as soon as their body temperature warms up to 54°F, so the ambient temperature can be considerably less if it's a sunny day. Now, bumble bees *will* fly in bad weather, but their prime season comes later in the year. In the early spring when the BOBs first come out, the only bumble bees alive and possibly out gathering are last year's queens-to-be, and there aren't going to be enough of them to pollinate an orchard.

Pollinating an orchard of early-spring fruit is something that BOBs do supremely well, completely outperforming honey bees on a bee-to-bee basis. Partly that's due to the BOBs' work ethic. They typically put in more hours per day than honey bees in the chill of early spring. They also carry their pollen differently than honey bees and bumble bees, transporting pollen dry in a brush of hairs underneath their bellies, rather than on their hind legs. This may not seem like the most efficient way to get the groceries home, but from a flower's point of view it's an excellent way to go about things. Dave Hunter says that BOBs belly flop into flowers. When those BOBs come flopping in, covered in easily-rubbed-off dry pollen, it's no doubt flower pollination nirvana.

Indeed, an acre of fruit trees that requires one or two honey bee hives (with 20,000 or more bees per hive) can be pollinated by 250 to 750 female BOBs. (The number of bees needed varies depending on the location, fruiting species, spacing, and size of the trees—and presumably the desperation of the grower.) That doesn't sound like many until you start adding up the acreage. Washington State produces about sixty percent of U.S. apples, a prime candidate for BOB pollination. In 2011 Washington had more than 167,000 acres of apple trees. At 250 to 750 females per acre, the state's apples would require somewhere between 41 million and 125 million female orchard mason bees. And then there are the cherries and the pears and the plums, not to mention all the other states.

The males don't get counted in this pollination game, because they aren't actively collecting pollen, although they do some incidental pollination when they visit flowers to feed. Nevertheless, the males are needed to help create the following year's generation and so must go into the orchards as well. Typically about fifteen to twenty male eggs are laid for every ten female eggs. So, all together, 60 million to 250 million BOBs would be needed to replace all the commercial honey bees in Washington's apple orchards.

Despite the obvious benefits of BOB pollination, most growers of early fruit rely on honey bee hives for their pollination services, hoping that the weather is good enough that the bees will deign to come out and forage rather than live off leftovers. Why don't they use BOBs instead, or even use BOBs as well as honey bees?

Honey bees are a known quantity. They have been managed for millennia and viewed as reliable pollinators for centuries. Not surprisingly, the knowledge base is smaller for raising and using BOBs than it is for honey bees. Also, if growers want to use BOBs, they would have to manage the bees themselves. Most growers don't keep honey bees; they rent them. The honey bees only come when needed and thankfully go away once they've done their job. With the beekeepers to look after the bees, the growers are able to focus on all the other parts of growing and selling fruit successfully. With the honey bees in so much trouble, however, various people are seeking alternatives and some of those alternatives involve BOBs.

BOB relatives: *Megachile*, leafcutting bees

Bees in the genus *Megachile* are called leafcutting bees because many use leaf parts to construct nest cells. A female *Megachile rotundata*, an alfalfa leafcutting bee, uses fourteen or fifteen leaf pieces to line just one cell. She uses her mandibles like scissors to cut each fragment precisely and flies it back to the nest, where she maneuvers it into place in the tight darkness. She nibbles the edges of the leaves so they'll be pulpy and stick together. The end result is a tiny leaf package containing one egg and all the food that egg will need to grow into an adult.

Megachile rotundata is a non-native bee, arriving in the United States in the 1930s, yet it has become one of the few widely used managed bees. They are an important pollinator of alfalfa seed crops (not the alfalfa used for sandwich sprouts, the kind used to grow hay to feed cows). The leaf packets can be popped out of their holes—they look a bit like tiny cigars—and stuck in a refrigerator until the bees are needed to pollinate the fields.

A female *Megachile perihirta*, the western leafcutting bee. All megachilid females—and only megachilid females—carry pollen on the underside of their abdomen. It is a distinguishing feature of the family.

A male *Megachile melanophaea* from Madison, Wisconsin. During sex, the male covers the female's eyes with those hairy front legs.

A *Megachile* leafcutting bee heading back to the nest with a piece of leaf. ◄

An *Osmia lignaria* male sporting a fine mustache.

Raising blue orchard bees

If I am a typical example, having gardeners raise BOBs may be a problem. When I went back to check on my little box of BOBs in the fridge a week or so after I'd stashed them, I found that the cotton ball that I'd hoped would act as a humidifier had turned into a hard, dry lump. And some of my cocoons were looking a bit shriveled. I resoaked the cotton ball, added some wet paper towels, put the box back in the fridge, and fretted. A few weeks later, when a fair number of plants had started to bloom, I decided it was time to stop mothering these bees. They were going out into the world. I put the bees, in their tiny box, on a little shelf in my bee house and, feeling relieved that my part was done, went on my way. Of course, like any mother I needed to check in, and when I did so a few days later all the cocoons were gone.

Stupefied, I looked around and started finding cocoons scattered all over: on a chair, on the deck, in a planter. Something had knocked my cocoons right out of their house. I never found all twenty cocoons; I assume some went between the cracks of the deck or possibly got eaten. One had the back of the cocoon ripped off, and a dead-looking bee lay inside. (It *had* been stuffed in there very tightly.) A few, a very few, appeared to have hatched, the cocoons open and empty. I returned the handful of unharmed cocoons to the bee house without much hope.

I'd tried the homeowner end of Dave's plan and run into some problems. I decided to visit an orchard grower who was using BOBs and see how they were working at that end.

I drive to Mosier, Oregon, located in a small valley an hour or so east of Portland. I am going to visit Kris Fade, a cherry and wine grape grower. It's mid-April, which is usually peak cherry blossom time in Mosier, but not this year. Seattle has experienced a warm spring with lots of flowers coming early and so, apparently, has Mosier. Kris had written me, "Orchard conditions have been really *crazy* this year." So, I'm not surprised to drive into a valley of rolling hills filled with the green of new leaves rather than the cherry blossom white I'd hoped for.

Kris and her husband's main pursuit is growing grapes and making wine, and I meet her at their tasting room. Kris looks to be in her thirties and is fashionable in a casual way, wearing cowboy boots and a plaid shirt with her honey-colored hair piled on her head. We head out into the orchard. It's a beautiful blue-sky day, and as Kris and I walk through row after row of Bing cherry trees, she tells me about raising and using BOBs. It is far from a precise science as far as a small grower is concerned.

Kris points out the newest iteration of bee hutches—covered homes for bee nest blocks—that she's developed after a couple of years of trial and error. It's a metal washtub turned on its side and attached to green metal fence posts so that the washtub is held about four to five feet off the ground. It's easy to take apart and store when the season is over. Inside each tub she puts a couple of nest blocks, which are grooved wooden trays banded together. Each groove is shaped like a half circle. One grooved tray faces up, the next faces down, and together they make holes. The bees feel all safe and cozy in their holes, not realizing that a person can pop the trays apart to expose all the babes. It's a neat system that allows the cocoons to easily be removed and cleaned come fall, when it is time to put the bees into cold storage. The trays can also be sanitized, removing any pests or diseases that might linger, and then be reused the following year.

BOB hutches at Kris Fade's cherry orchard. ▲

A BOB heading home to a wooden tray nest block in Kris's orchard.

Behind each hutch is a hole in the ground with water dripping into it, making the mud these mason bees need to construct the walls between the nest cells for their babes. Not all wet dirt will work for mason bees; it has to stick together. Really sandy soils won't do the job of making cell walls. To be successful with BOBs you need only a few things, but each is essential: proper sized holes (about five-sixteenths of an inch), which may be anything from paper tubes to reeds to drilled wooden blocks to beetle burrows; appropriate early-season flowers that are rich in pollen and nectar; and mud. If any of these three are missing, the BOBs won't prosper.

Savior bees?

The life of a BOB is pretty simple. They are solitary bees, living their short adult lives alone. The males mate, sleep, eat, and eventually die. The females mate and spend the rest of their lives busily provisioning nest cells and laying eggs (with some sleeping and eating as well), but they do all of this alone.

A normal day for a BOB female goes something like this. She finds a fine-looking hole and goes off to get some mud. She gathers the mud with her mandibles (part of her elaborate mouthparts) and shapes it into a tidy ball. She wedges the ball between her mandibles and front legs and flies it back home. Her first mud wall is built to seal up the back of the nest, and it takes about ten trips to the mud pit to make that one partition. Assuming the mud supply is reasonably close, the trips don't take too long, a minute or two each. Once the wall is built, she goes off to start collecting pollen and nectar for her first babe.

While she's slurping up nectar (which she stores in her crop), she uses her legs to scrabble up pollen from the anthers. She stashes the pollen on the hairs on the underside of her belly. A female BOB visits around seventy-five flowers on each trip, and it usually takes about ten to fifteen minutes. When you factor in flying time, this likely gives her only a few seconds per flower. This is clearly no leisurely jaunt to the grocery store; it's more like careening down the aisles madly tossing food into the grocery cart.

Once the BOB female is loaded up, she goes back to her hole, climbs in, and spits up the nectar. Now, a problem arises: the pollen is on her back end and the hole is small, with no room to do a somersault. So she backs out, turns around, and backs in so that she can offload the pollen. Then she's off for another run.

It takes between fifteen and thirty-five trips to get enough food for one cell, meaning a BOB female is likely to visit between 1125 and 2625 flowers to get enough food for just one babe. In commercial orchards, a BOB usually completes around seven to thirteen nest cells. So, to provide food for all those babes she must visit at least 7875 flowers and maybe as many

Cavity nesters: *Hylaeus*, the yellow-faced bee

All the bees in the genus *Hylaeus* are tiny. They nest in premade holes, mostly aboveground, and in that way they are like BOBs. Researchers studied the nesting habit of one of these bees, *Hylaeus bisinuatus*, in a greenhouse, using glass tubes, so they could see exactly what was going on.

To begin a nest cell, a female secretes a viscous material—silk embedded in a clear polyester—that she spreads around with her tongue to coat the cell walls. She makes twenty to fifty parallel strips along the lower cell wall before moving to a new spot and spreading more. When she has completely coated her baby room in silk and polyester, she flies off for provisions.

Hylaeus bees are unusual in that they carry pollen (mixed with nectar) internally in their crop, a useful addition to the digestive tract that can be used for temporary food storage. The female spits up this liquidy mix into her carefully coated nest cell, then heads back out for more. When she has gathered enough food she lays an egg, right in the provisions, where it floats, partially submerged. She then closes up that cell and begins the next.

A teeny *Hylaeus* alights on a flower in South Carolina.

as 34,125. That last number must count as a really bad year. It all sounds very tiring and frantic. It's doesn't seem surprising that each female BOB only lives about twenty days as an adult, and by the end of that time she's moving slowly, her hair worn off and her wings tattered.

However good BOBs may be individually as pollinators, they seem like unlikely potential saviors of the honey bees. Tens of thousands of honey bees live together in a box that can be conveniently carted around to fields and orchards when needed for their pollination services. BOBs live alone and lay a couple of eggs in a hole but they are willing to live in man-made holes right next to each other. When a bunch of bees nest close together they are said to be gregarious nesters, and, although all BOBs don't live this way, they are willing to do so. This willingness makes Dave's savior plan possible.

A lot of BOB-sized holes can be packed into a surprisingly small space. One BOB house from Crown Bees, which is fourteen inches tall, ten inches wide, and nine inches deep, holds 200 nesting tubes. Assuming six cocoons per tube, that's potentially 1200 mason bee cocoons in a vertical surface no bigger than a legal pad. Once you've got the bees in a block or a bunch of paper tubes or reeds, significant numbers can be moved to where they're needed, just like the honey bees.

The BOBs' willingness to nest gregariously and their easy portability are why Dave's plan might work. He sells orchard mason bees to gardeners like me, usually ten to twenty cocoons. Over a few seasons their numbers will hopefully increase naturally until they pass 200; that number, Dave says, is adequate for pretty much any yard. Then people can sell (or trade for supplies) their surplus cocoons back to him. Dave, in turn, will sell the bees to orchard growers to supplement, or possibly replace, the services of honey bees for early-season crops.

Back in the cherry orchards

As we continue ambling along, Kris points out blue boxes in some of the trees. These were her first bee hutches. She'd been told to tie these boxes

Another BOB relative: *Anthidium*, the wool carder bee

Bees in the genus *Anthidium* are called wool carder bees because the females use their sharp mandibles to shave the fuzz off of wooly-leaved plants like *Stachys byzantina* (lamb's ear). The female rolls the fuzz up into a ball and carries it back to her nest. She uses the fuzz to shape a fluffy pillow that will hold a single egg and all the pollen and nectar the babe will need to grow to adulthood.

One species commonly seen in gardens is an interloper from Europe, *Anthidium manicatum*. These bees are honey bee–sized, and each bee has bright yellow stripes atop its abdomen (not yellow hairs, but stripes on the exoskeleton itself). These wool carder bees seem to have a lot of sex or are exhibitionists, because I see more photos of them mating than I do of other bees. The males of this species are aggressively territorial. They patrol flowers to guard them and any visiting females from intruders of any sort. They will pounce and wrestle down all manner of flying insect invaders.

An *Anthidium manicatum* male on a territorial perch in Sacramento, California.

I saw a paper about this bee entitled, "A Character Analysis of a Solitary Bee, *Hoplitis albifrons* (Hymenoptera, Megachilidae)." Who knew bees might have character flaws in need of analysis?

into the trees and put the BOB nest blocks inside to keep them dry. The big problem with this original housing method revealed itself when it was time to spray the trees. Cherries are prone to all sorts of pests and diseases, and, although Kris is an organic farmer, she still sprays, often with sulfur, a fungicide allowed on organic farms. Of course, Kris tells me, she can't spray during the day when the bees are out. She has to wait until after dark, when the bees have gone to bed. In previous years, on spraying days Kris would strap on a head lamp and head out into the orchard with a pile of old-fashioned plastic shower caps. "We had to shower cap every single box," Kris says, to keep the spray off the bees. The boxes weren't on every tree so she'd be "stumbling around in the dark" trying to find the blue boxes among the thousands of cherry trees. Once she was done, she could go to sleep, and her foreman would come out around midnight to spray. Then she'd be up again at sunrise to take off all the wet, smelly, sulfur-coated caps before the bees got up. But "this year," she says, "I'm not doing that."

Instead, she came up with those washtub homes that she put in the rows between the trees. Kris still has to go out at night to cover the bees before spraying, but the washtubs are easy to find and she can just throw a garbage bag over them.

So, in three years, Kris figured out the best way—for herself and for the bees—to manage the nest blocks. Problems continue to arise, but Kris plans to stick with the BOBs. She says that she wasn't able to manage side-by-side testing of the parts of the orchard using BOBs to those without, but her overall yields are up. Plus, she says, the areas with the BOBs just look like they have higher yields. Several research studies support her impression, showing that BOBs (and some of their close relatives) do indeed bring in the cherries.

In 1997, a cherry grower along northern Utah's Fruit Way went to the scientists at the Bee Lab in Logan looking for BOBs. The grower hadn't been satisfied with what the honey bees were doing for him. Plus, some neighbors in a new nearby subdivision had complained about stings. He'd read about BOBs and was hoping to find some. The folks at the Bee Lab recognized a research opportunity when it came calling, so they gathered some of their own BOBs and took them out to the cherry grower's orchard.

For five years they put BOBs in that orchard and tracked what happened. The researchers looked at things like cherry yields, how many BOBs successfully nested, and the weather. They compared the results to the previous four years, when honey bees had been on the job. They found that the weather was essentially the same, but the yields in years good enough for cherries to be harvested ran about two times greater with the BOBs. Furthermore, in two bad-weather years the grower with BOBs was able to harvest some cherries, while many of his neighbors who used honey bees couldn't. His cherries commanded top dollar. BOBs got the job done.

Interest in BOBs is growing. Kris says that other growers in Mosier are considering using them. And then there's Gordon Wardell, the Wonderful Company's head of pollination, with his twenty acres of net houses filled with beautiful blue flowers and busy BOBs. He's investigating the best way to raise monumental numbers of BOBs in a small area. What are the best plants, the best hole size, the best hutch locations? He's got a huge walk-in temperature-controlled unit so that he can wake those BOBs up

All *Osmia* species aren't created equal

Osmia lignaria is only one of around 135 species of *Osmia* in North America, and those other *Osmia* species are surprisingly different from BOBs. They tend toward the same mesomorphic build, but they come in different lengths and colors. Their nesting habits range from the normal to the little bit odd: rose stems, beetle burrows, shallow holes underground, mud homes aboveground, and even dry cow patties. Some use plain mud for partitions, while others use balls of leaf pulp rolled in mud. A few even choose to live in snail shells. In fact, the use of snail shells is so common for some European *Osmia* species that scientists have a name for them: helicophyle *Osmia*. Different species treat their snail houses differently. One closes off her house with chewed-up strawberry leaves, and another uses rabbit or sheep dung. A versatile and pragmatic group, these *Osmia* bees.

A glorious *Osmia aglaia* female visiting *Penstemon heterophyllus* in Sacramento, California.

early, in time for the almond bloom. He's hoping to get a million bees from his twenty acres of net houses. One million bees. I'm pretty sure he's not covering up his nest blocks with shower caps.

Harvesting blue orchard bee cocoons

I had given up hope for my BOBs after all the problems I'd visited upon them, but when I went out one late spring day and looked up at my bee house, I saw two tubes closed up with daubs of mud. Yes! I had bees. By mid June I had five tubes filled. Were any from my original bees or were there other BOBs already resident in the area just happy to see a home? According to Jordi Bosch and Bill Kemp in their handbook *How to Manage the Blue Orchard Bee as an Orchard Pollinator*, BOBs do like to go back to the same nest area they came from, making it more likely these were some of my original bees. I paid around $35 (with taxes and shipping) for my twenty cocoons. If I got six eggs per tube and they all survive, that's thirty whole bees, probably about ten of them females. A little over a dollar a bee, which seems a bit pricey. Although when viewed as a fifty percent increase over my starting numbers, that's not bad, given what I had put the bees through.

I go to the fall harvesting party at Crown Bees, where I meet other people who have done much better. Harvesting is just removing bees from their holes and sanitizing the cocoons before storing them for the winter. It's important to wait until autumn comes to do this. Of course, bees can be left in their holes, but Dave Hunter strongly recommends the sanitizing bit to keep the disease and parasite loads down and increase the chance of bee survival. I usually choose the least work option available, and at the moment I'm willing to accept the bee loss that will no doubt come with letting the bees do it the wild way. That said, it is fun to see what is inside those holes.

I see maybe two dozen people while I'm at the harvesting party. I help one woman who has thousands of bees to harvest. We peel open paper tubes and dump the contents—cocoons, mud, pollen, frass (bee poop), and anything else—into an aluminum roasting pan. Most of her cocoons look good, few have the BOB problems typical for the Seattle area. Pollen mites had gotten into a few of the cells, leaving behind a pile of what looked like dried yellow pollen where the cocoon should have been. We find only one case of chalkbrood, a fungus. Chalkbrood attacks larvae and goes through an array of color changes until it reaches the nasty, black, spore-producing phase, when you pick up the former larva and it bursts into a batch of black powder, ready to spread. Fortunately, she has no monos, tiny parasitic wasps. The genus name for these wasps is *Monodontomerus*, so I understand why people just call them monos.

In southeastern Washington, some bees are actually managed in the ground. I walked across this man-made alkali flat while thousands of alkali bees, *Nomia melanderi*, swirled about below my knees, unbothered by my presence. Alkali bees are excellent pollinators of alfalfa seed crops.

After getting all the cocoons out of their holes, the next step in the sanitizing process is to dump the contents of the roasting pan onto a sieve to get rid of all the extraneous bits. Then the cocoons go into a big vat of a bleach-water mix. The cocoons float, swirling and bumping into each other as they are stirred around for a minute of so. The agitation is supposed to remove the pollen mites and the bleach the chalkbrood spores. The cocoons are rinsed, and once dry, can be stored in a fridge (or other cool place) until it's pollination season again.

Are managed bees the way to go?

Dave's plan for BOBs introduces people to a different view of bees, but is it really an opportunity to help honey bees and safeguard our food supply? Managed honey bees and bumble bees both have experienced problems arising from factors associated with that management. Another managed bee, the alfalfa leafcutting bee (*Megachile rotundata*), also has management problems. Attempts to multiply the bees in the United States have generally come up short, while efforts in the Canadian prairie provinces have seen regular increases. Might large-scale efforts to domesticate BOBs also lead to problems? North America has 135 species of *Osmia*. Could diseases and parasites be spread from managed BOB populations to their many relatives? Certainly Dave tries to prevent the spread of any problems, sanitizing cocoons and only sending bees back to the locality they came from. Cocoons sent to him from Washington State only get sold to folks in Washington. Of course, someone else might make different decisions, or Dave's efforts might not be enough.

In addition to worries about spreading diseases and pests, managed bees can impact wild bees in other unexpected ways. Beekeepers seeking places with abundant, pesticide-free floral resources for their honey bees during the summer look to public lands in the western states: land and flowers already being used by local wild bees. What happens when a bunch of honey bee hives gets dropped down in their midst? A recent study estimated that just one honey bee colony collects 650,000 pollen pellets over three months of summer pasturage. That's enough pollen to feed around 110,000 babes of an average-sized solitary bee. A typical beekeeper might have forty hives, and some have many more. That adds up to a whopping pollen loss. The few studies that have been done suggest that there is likely not enough surplus pollen to support both the local wild bees and the newly arrived ravening horde of honey bees.

So, do we try and find yet more bees to manage, not knowing what unexpected repercussions might result, or should we be focusing on another method—bringing wild bees back to the farm? Clearly, once upon a time pollination was done largely by wild bees. What changed? I head to Maine, where a native wild crop is king and where 85,000 honey bee hives are trucked in during bloom time. What happened to all the wild bees? I thought pesticides might be the problem, but it turns out the answer isn't that simple.

Bees, Blueberries, Budworms, and Pesticides

THE BATTLE OF the budworm began in 1952 at Budworm City in the upper reaches of New Brunswick, Canada. Given that the entire place consisted of sixteen shacks and an air strip buried in the woods, Budworm Village would have been a more appropriate name. Nevertheless, from that air strip, pilots flew World War II surplus biplanes loaded down with the chemicals needed for the aerial assault on the enemy, *Choristoneura fumiferana*, a small gray-and-brown moth. The innocuous-looking adult wasn't the true problem; it was the ravenous young larva, a caterpillar called the eastern spruce budworm, that they sought to kill.

This spruce budworm is native to northeastern North America, and its effect on the conifer forests there is calamitous. The budworm overwinters in its larval form, rousing from its winter diapause in the spring, when it begins munching the leaf bases and buds of certain conifers. Its favorite food is the balsam fir (*Abies balsamea*), but it's reasonably content to eat several other conifer species as well. By summer, the chewed needles have turned reddish brown and begin to fall, leaving skeletal gray branches. Some trees may die in a year, and not many can last more than a few years of repeated defoliation.

The budworms are always present, averaging about five per tree, but records going back to the 1700s show massive outbreaks occurring on average every thirty-five to forty years. During an outbreak, budworm numbers explode to as many as 20,000 per tree. Each outbreak typically lasts from about three to ten years. Trees die, fire often follows, and the forest is renewed. The devastation is completely natural.

The problem started when humans wanted the same trees for the pulp and paper mills that the budworms wanted for dinner. The pulp and paper industry's business model didn't include allowances for multiple years off due to massive tree die-offs. So, when the early signs of a new outbreak began in New Brunswick in the 1940s the pulp and paper folks got worried. They started a study into the problem, and the entomologist Reginald Balch suggested spraying the forests with chemical pesticides from airplanes, a relatively new idea at that time. The plan was approved, and Budworm City was built in 1951 by the New Brunswick International Paper Company as the launch pad for the spraying program.

The pesticide chosen was the wonder chemical of the time, DDT. The first sprayings were deemed incredibly effective, killing ninety-nine percent of the budworms. Yet the budworm epidemic continued. The New Brunswick provincial government and four pulp and paper companies formed a partnership named Forest Protection Ltd. (FPL) to continue spraying the forests of New Brunswick. Five years after those sixteen planes dumped the first loads of DDT, the program had grown to 200 planes spraying almost 5 million acres of land. Between 1952 and 1968 more than 17 million gallons of pesticide were dropped over 28 million acres of New Brunswick; most of it was DDT.

Protests against the spraying began first with fishing and hunting groups, but the scientists got into it too. Rachel Carson's *Silent Spring* came out in 1962, showing the dark side of DDT to a wide audience. Eventually, FPL agreed to stop spraying DDT but the spruce budworm outbreak was still going, having lasted much longer than usual. The efforts to save at least some of the forest so it could be harvested meant the budworms never ran out of food. Even a ninety-nine percent kill rate leaves 200 budworms per tree (assuming that 20,000 budworm starting point), plenty of budworms to keep the outbreak alive, especially if a lot of the budworm's natural predators were killed by the pesticide as well. In 1969 FPL switched from DDT to fenitrothion, and an unexpected group of animals started dying—bees—and with them went another surprising victim—wild blueberries.

Bees and wild blueberries

A wild blueberry field doesn't look like a place where food is growing. It's beautiful. Green is the dominant color of the field, even in flower, but the shades of green shift and mutate: darker here, lighter over there, a hint of red in the distance. On the day I visit in late May, little pools of bell-shaped flowers waver above the green. Most of the flowers are white, but some glimmer a pale sweet pink and a few are white with pink stripes. It's like a master quilter has come along and pieced together the hillside.

I've come to Maine to talk to Frank Drummond of the University of Maine at Orono about wild blueberries and their pollinators. Our first stop is Highland Blueberry Farm, a small organic farm not too far from the coast, west of Acadia National Park and east of Bangor (Maine's third most populous city, population around 33,000). Frank leads me out into a field. Small, springy, thickly set stems rise straight up from the ground. The stems aren't individual little plants. Instead they arise from underground stems (rhizomes), so each plant, called a clone, is several feet across. Each clone is genetically distinct, accounting for the pretty patchwork effect. The field has no rows or paths, and I look for bare spots to put my feet. The plants are low, much lower than I had expected, only ankle high—harvesting them by hand must be hideous—and it's hard not to step on them. Frank notices me mincing along and tells me not to worry; I'm not going to hurt the blueberries. He did studies on it years ago to check. I try to walk naturally after that, but it still feels wrong to just stomp around on somebody's crop plants.

These aren't the blueberries that we buy fresh in the grocery store; those come from cultivars of highbush blueberries, *Vaccinium corymbosum*. A cultivar differs from a wild plant in that it has been selected for certain desirable traits, and those traits can be maintained when the plant is propagated. Cultivars of any kind of plant get names—Bluecrop

Honey bee hives tend to get dropped in large groups in the blueberry barrens of Maine.

blueberries, Sungold tomatoes, Pink Lady apples, Lollo Rossa lettuce. The propagation may be done in a host of ways, ranging from old-fashioned seeds to high-tech tissue culture. Wild plants, like the lowbush blueberries, just grow, the genetics a random lottery allotted by the bees.

The importance of wild blueberries for the wild bees of Maine can scarcely be overstated. The current count for bee species in Maine runs to 276, and about half of them have been found visiting the little blueberry flowers, albeit not all at once in one field.

It is the usual two-way pollination street, because the blueberries truly need the bees for pollination to occur. The blueberries can spread along nicely by their rhizomes for dozens of years, but to make brand new plants, to mix up their genomes, they need bees. Wind is not an adequate pollinating force for blueberries. Lowbush blueberries have fairly heavy, sticky pollen and the flowers open downward with the stigma just peeking out the end of the little white cup. Some strange, strong ankle-high wind blowing pollen around at ground level might get a few grains stuck to a stigma, but not very many. Furthermore, most blueberry plants are self-incompatible,

A male *Augochlorella aurata*, a common visitor to Maine's lowbush blueberry fields, with his mouthparts extended. It's not clear what he's after on this leaf.

so they need pollen from a genetically different plant to reproduce.

Ease of pollination is further hampered by how lowbush blueberries hold their pollen. Like tomatoes, they have poricidal anthers that require shaking to get the pollen out. Bumble bees use their flight muscles to buzz pollinate blueberries, just like they do with tomatoes, and *Andrena* mining bees use the same method. Other bees might not shiver their flight muscles to shake pollen out, but they've found other ways. The Maine blueberry bee, *Osmia atriventris*, drums the anthers with its forelegs. (I didn't get to see this while I was there. But when I try to picture a bee stuck up inside a small flower, flailing away, my imaginary bee is also yodeling. Perhaps each act seems equally improbable?) Honey bees are no better at shaking pollen out of blueberry flowers than they are with tomatoes, but they do visit the flowers to get nectar and so do some incidental pollination.

Lasioglossum cressonii isn't a showy bee but it can be plentiful. In a three-year study of Maine lowbush blueberry visitors, this species was the overall winner in terms of abundance in two of the years. ▲

This *Osmia atriventris* female looks like she's punked out her hair. This bee drums the anthers to get the pollen out. ◄

Bombus ternarius, the tri-colored bumble bee, was the bumble bee winner in terms of numbers in the three-year study of lowbush blueberry visitors. ▲

Halictus confusus is another visitor to Maine blueberry fields. ◄

Burning blueberries

Frank Drummond takes me to meet the farm's owner, Theresa Gaffney, and we all head uphill from the pretty patchwork field to another—this one bare and blackened. Wild blueberries run on a two-year cycle: harvest one year, burn or mow to the ground the next. The burning or mowing gets rid of old growth, forcing brand new shoots to come up from the rhizomes. First-year shoots produce more flowers and hence have the potential to yield more fruit. Those new shoots are also easier to hand-harvest than older ones. (Harvesting lowbush blueberries by hand actually means using a rake, not picking the berries with your hands.)

Burning has historically been an essential part of lowbush blueberry care. Now, many growers mow rather than burn to save money. The method used for burning varies, but all are expensive. One method used a Woolery blueberry burner that I was told "spewed flames out the back like a constipated dragon." (Those were the days.) The switch away from burning has had a negative impact on bees. Burning practices started changing around 1930, when some growers decided to burn huge swathes at once rather than doing a patchwork of small burns. Suddenly, there were fewer bees. Many bees like to nest in the burned ground, but they will only fly so far from nest sites to flowers. If the burned area is too far from the blueberry flowers, there's a problem.

Burns are useful for the native bees in other ways. Frank points to an *Andrena* mining bee sitting quietly on the blackened earth. He explains that she's sitting there because, unlike a bumble bee queen, she can't control her internal temperature and with her small size she loses heat fast. On chilly spring days, these dark burned spots and the rocks in the field absorb and radiate heat, allowing little bees to get out and forage, while stopping to take periodic warm-up breaks.

Not many bees are out on the day I visit, probably because the wind is up. This is both good and bad. The wind keeps the biting black flies, Maine's "Defenders of the Wilderness," at bay, but it also keeps some of

Colletes, cellophane bees

Mating practices vary among bees: the length, the location, the frequency. Many of today's scientific papers are clinical, but not those written in the early twentieth century by Phil and Nellie Rau. Of *Colletes compactus* they wrote, "There at the side of the water was a swarm of *Colletes* bees, perhaps two hundred in number, buzzing, flying, wheeling, dancing, weaving in and out. . . . The excitement was riotous, dancing and mating, dancing and mating."

S.W.T. Batra wrote more prosaically of a *Colletes inaequalis* male just after the ground had thawed. The male was "grasping" the female but was "too cold . . . to move or copulate." I can only imagine what the Raus would have written about that desperate male.

Colletes inaequalis visits Maine's lowbush blueberries. Members of this genus are called cellophane bees because the female secretes a waterproof coating that she spreads onto the walls of her nest cell with her tongue.

This *Andrena* female from Pickens, South Carolina, guards her burrow.

the bees at home. Nevertheless, as Theresa tells me about blueberry farming, Frank is casting about, hunting for bees.

He finds some and calls out, "You can see a couple of *Andrena* flying close to the surface in a zigzag way, looking for females." Frank explains that these males are likely to be out of luck. Most of the females mated with the males that had their act together back when the willows were in bloom. Of course, those guys are likely long dead.

"So these are the last dregs?" I ask.

"The last hope to incorporate their gene pool," Frank agrees.

Hi ho, hi ho, it's off to work we go

Many species of *Andrena* are common out in the blueberry fields. They are called mining bees and get that name from their method of nesting. Back in the 1970s, Martha Schrader and Wallace LaBerge studied *Andrena* bees in lowbush blueberry fields in Maine and New Brunswick. They wrote a description of *Andrena regularis* starting a nest. The bee flies up loaded with pollen, but for some reason she has not already set up her nest, so

In each year of a three-year study of the Maine lowbush blueberry fields, *Andrena carlini* was the most abundant member of the genus found. ▲

An *Andrena frigida* male with a large, bushy, positively Victorian-era mustache. ◀

she begins digging. Dissatisfied, she soon quits and starts a new hole. For whatever reason the second spot is better than the first, and she digs vigorously, shoving bits of dirt out from underneath her body. She crunches the larger clumps of soil into submission with her mandibles. Pretty soon she is out of sight down the hole, with bits of dirt occasionally getting ejected from the entrance. After about twenty minutes she comes out, flies a zigzag orienting flight, and then zips back into her hole to dig some more. *Andrena* bees are called miners for a reason.

Ground nesting is the norm among bees, with about seventy percent setting up their household somewhere underground. Some may use preexisting holes such as old rodent holes; others excavate their own. How bees set up their nest cells varies. Some place the cells around the main tunnel like the spokes on a wheel, while others have cells that are seemingly randomly arranged in side tunnels. Some burrow deep, and others stay close to the surface. The species that Schrader and LaBerge studied dug out its nests to what seems like a rather paltry depth of six to seven inches, but an *Andrena regularis* is only about half an inch long. That bee was digging a tunnel twelve to fifteen body lengths deep. That's like a six-foot-tall man digging a tunnel that's seventy-two to ninety feet deep. From a bee's point of view, these really are mine shafts. In other areas these bees have done deep mining, digging down to a whopping depth of eighteen inches.

Andrena bees often nest in aggregations, sometimes with the nest entrances exceedingly close together. One aggregation in an abandoned blueberry field in New Brunswick had thirty-six nest entrances in a four foot by three foot area. Some of the entrance holes were only an inch apart. These miners may be solitary bees, but they certainly don't mind being close to their neighbors. They aren't exactly apartment dwellers since they all have their own front door; it's more like living in townhomes.

Some species of *Andrena* do live like apartment dwellers. These are the communal nesters, where up to a few dozen bees share one front door. Once in the ground, each bee digs her own side tunnel and provisions her own nest. One nest of *Andrena crataegi* in Maine had forty-four

A sniff test for *Andrena*?

Telling one *Andrena* species from another is notoriously difficult for beginners. Even recognizing that a bee is some sort of *Andrena* can be difficult. I thought I might have found a fun identifier when I ran across a paper entitled, "Comparative Analyses of Lemon-Smelling Secretions from Heads of *Andrena* F. (Hymenoptera, Apoidea)." Lemon-fresh bees? Regretfully, lemons aren't the only thing *Andrena* bees may smell like. Robbin Thorp mentions two other groups of smells: a musky oniony odor and a flowery sweet one. Some other bees have fragrances as well, so I've given up on the idea of using an easy sniff test for *Andrena*.

A little *Andrena* bee in a magnolia flower.

females using the same nest entrance. Apparently, the morning exodus was orderly, with the bees leaving one to two minutes apart and no Keystone Kops moments of two bees trying to get out at once.

Andrena species are common visitors to wild blueberries all over Maine and southeastern Canada. Thirty-one different species have been found visiting blueberries in Maine. When the fenitrothion rained down in New Brunswick, *Andrena* took a hit, and they weren't alone.

A short history of pesticides

Just about everyone practices pest management. We swat mosquitoes, and it's a quick death to any slug seen snarfling up the lettuce. The extent and methods people devote to pest management differ, but to manage significant numbers of a particular pest, one usually has to resort to some sort of pesticide. A pesticide is a substance used to kill, prevent, or repel a pest, whatever that pest might be. All sorts of things qualify as pests—rodents, fungi, weeds, insects, mites, and many more—and each group has their special -cides for offing them. Clearly, knowing what you're trying to kill is critical for using the right stuff. You're likely to be deeply disappointed if you try to kill those little pests with an insecticide when they are actually mites, which are relatives of spiders and not insects at all.

In 2007 (the most recent year for which I could find data), Americans spent more than $12 billion on various kinds of pesticides. A little over a fifth of that spending was for home and garden use. About two-thirds was spent for agricultural use, and the rest was spent by the government, industry, and other commercial interests. In that same year, Americans used more than 850 million pounds of pesticide active ingredients, with 66 million of those pounds being used in our homes and gardens. We may be taking pesticide use to new heights, but people have been dealing wholesale death to pests to the best of their ability for millennia.

In 2500 B.C., the Sumerians reported rubbing foul-smelling sulfur compounds on themselves, hoping the stench would drive off personal pests. The Egyptians listed more than 800 recipes used as pesticides and poisons in the Ebers' Papyrus, the oldest known medical document. The ancient Chinese used predatory ants to protect their citrus orchards from caterpillars and beetles. They even provided bamboo bridges between branches to make it easier for the ants to make their patrols. The Romans were troubled by rust (a kind of fungus that indeed looks rusty) on their grains, but they tried a different route to control, making offerings to the rust goddess, Robigo. The Romans didn't just rely on random gods, though; they invented the first known chemical weed killer, amurca, in the first century B.C.

During the Middle Ages, pest control experimentation languished, at least in Europe. One article suggested that the blights and pest outbreaks of the Middle Ages were seen as rightful retributions from God and so shouldn't be fought. The Renaissance brought an end to that sort of defeatist attitude. Tobacco was used to control lace bugs on pears during the seventeenth century. More pesticides kept coming along, and a sprayer was developed in 1880 so the pesticides could be delivered as a fine mist. The first aerial spraying came in 1921, when lead-arsenate dust was dumped on catalpa sphinx moths in Ohio. A lot of these pesticides were not terribly effective, but something much better was coming along from a surprising source, coal tar, also known as creosote.

Coal tar is a by-product of the coal coking process. Coal can be burnt as is to provide energy, but coking improves it. Heating coal up to 1800–3600°F (980–1960°C) in an oxygen-free environment drives off volatiles: water, coal gas, and coal tar. The material left behind, aptly named coke, is a purified, high-carbon material that is excellent for making steel and burns without producing much smoke.

Nineteenth-century Londoners found a use for one of the by-products, the coal gas; they used it to light the city. At first, the coal tar was seen as garbage, and it got dumped in the Thames or buried in the countryside. Eventually, it was used in shipbuilding and wood preservation. The true value of coal tar wasn't discovered until people started looking at the many carbon compounds that make up that black viscous substance. They found organic compounds that were the starting point for a whole array of useful items: synthetic medicines, scents, flavorings, fertilizers, explosives, dyes, and pesticides.

In 1935 Paul Müller, a chemist and physicist who worked for the Swiss company Geigy, was given the job of developing a synthetic insecticide. He wanted the insecticide to kill lots of arthropods quickly with little harm to mammals or plants, to smell inoffensive and be non-irritating to humans, to be persistent, and to be inexpensive to produce. Müller's team hunted and synthesized, and in 1939 they came up with a substance that met their requirements: DDT. (The compound had been created before, in 1873, but its insecticidal properties hadn't been recognized at the time.)

DDT was a truly miraculous substance. In one of his experiments, Müller noted that his fly cage was so deadly that even after he cleaned it any fly that touched the wall fell to the floor. He had to wait a month before the cages weren't death traps. A miracle substance indeed.

DDT wiped out mosquito-borne malaria in several countries and stopped epidemic typhus, which is transmitted by insects like fleas and lice, in Italy during World War II. DDT destroyed crop pests, leading to increased yields. However, it also kills fish and other aquatic life. DDT bioaccumulates as it goes up the food chain, becoming concentrated in the bodies of top predators. One of DDT's breakdown products, DDE, thins the shells of bird eggs, causing them to break while being incubated. DDT's toxicity to bees is surprisingly low, but it was just the first of many insecticides to come.

Pesticide risk to bees

When FPL, the consortium of business and government interests that had been blanketing New Brunswick in DDT, switched to fenitrothion, life got better for some animals and worse for others, particularly for bees.

Now, all companies hoping to sell a pesticide have to supply a vast amount of information to whatever government agencies regulate pesticides in that area. To sell in California, for example, a company must get approval from the EPA and the State of California. One piece of information the hopeful pesticide makers have to supply is what the pesticide does to non-target organisms, those that are just random passersby. Historically, the focus was on risks to people and other vertebrates. Bees came pretty far down the priority list.

Assessing the risk to these non-target groups is done by testing the product on them. Among the primary things the agencies want to find out is the acute toxicity, that is, how much of the pesticide it takes to kill the animal in a short period of time. They look at how much of the stuff could either be put on (contact toxicity) or fed to (oral toxicity) the animal before half the animals treated that way die. This is the LD_{50} (lethal dose, fifty percent). At the same time, the agencies are keeping an eye out for other, sublethal, effects.

For bees, the agency researchers used honey bee workers to determine acute toxicity, but they also looked at residual toxicity. Many insecticides are added to water and sprayed on plants. Any target insects hanging out (and probably munching away) when the spraying occurs die from contact with the insecticide. Or, after spraying, a plant-eating insect might fly up and take a little chomp out of a leaf coated in insecticide and die. Sometimes, even after the insecticide has dried, it's still toxic enough that when an insect just walks on it, the critter dies. This is the residual toxicity. So, bees are non-target organisms that might die from eating the insecticide when gathering nectar or pollen that was coated in the stuff, by being drenched when the spray came down, or by walking on a dry but recently sprayed plant.

In all these toxicity experiments, honey bee workers were the proxy for all bees. Until recently, the researchers didn't look into what non-lethal doses might do to bees. Also, all tests were done only on adult worker honey bees. What effects they might have on eggs, larvae, pupae, drones, or the queen were ignored.

The amount of insecticide involved can be quite tiny. The LD_{50} is usually given in either parts per million or parts per billion. For a little perspective, one part per billion is one microgram (1/1,000,000 of a gram) in a kilogram. One second out of about 32 years would be equivalent to a part per billion. The LD_{50} for bees (since they weigh much less than a kilogram) is measured in how many micrograms it takes to kill a bee; so basically it is the dose per bee. The LD_{50} numbers I found have fenitrothion being 200 times more toxic to a bee than DDT is.

Risk is not just about toxicity, though. It's about toxicity *and* exposure. Despite the fact that we're talking only a couple of parts per billion per bee, DDT wasn't considered particularly harmful to bees because they just didn't pick up those amounts out in the environment. The same wasn't true for fenitrothion.

The battle of the budworm continues

When FPL started using fenitrothion in New Brunswick, things got worse for bees. There was enough fenitrothion out there to do harm, and the blueberry growers with fields adjacent to sprayed forest areas saw it.

Peter Kevan was a young post-doctoral fellow when he got the call from Bridges Brothers Ltd. Blueberry Farms. The owners thought that the fenitrothion sprayings were killing the bees that pollinated their blueberries, and they wanted someone to come see if it was true. So, in 1972 Kevan headed out to New Brunswick and caught bees in blueberry fields both within the sprayed zone and outside it.

It turns out the growers were right; the fields adjacent to the sprayed zones had significantly fewer bees. In an article he coauthored in 1974, Kevan said, "This decimation of pollinators was so marked that merely standing on the edge of the fields the lack of pollinators could be appreciated by ear—the low buzz of a busy field was gone." Dead bee carcasses were also sent off for testing, and some showed fenitrothion present at much higher levels than the LD_{50}. Bridges Brothers sued FPL in civil court asking for an injunction to cease spraying plus $1.5 million for losses for the years 1970, 1971, and 1972. Kevan was an expert witness at the trial.

Decades later when I talked to Peter Kevan about that time he told me, "The scientific evidence that fenitrothion was doing a job on the wild blueberry pollinators resulted in a fair amount of rather acrimonious mudslinging." Particularly, he said, from the governmental forestry folks and the group doing the spraying, FPL. The planes weren't supposed to be spraying the blueberry fields, of course, but insecticide was definitely landing on them. Special paper that changes color when it gets insecticide on it was put out in the blueberry fields, and the color did indeed change. Those bee carcasses loaded with fenitrothion were additional proof.

Neonicotinoids and new guidance on risk assessments for bees

The old way of assessing risk to bees—looking only at what kills adult worker honey bees—has changed, largely because of a relatively new class of pesticides, the neonicotinoids. Historically, insecticides were placed on the outside of plants, and many still are. These pesticides get broken down by the sun and washed off by the rain. Neonicotinoid pesticides are different, because they are taken up *inside* the cells of the plant and can be translocated to new parts of the plant as it grows. Tidy. Coat a seed, protect a plant.

It seems like a great idea because the insecticide is always there. But it gets into the nectar and pollen as well as the leaves, which opened the door on all sorts of problems that people hadn't considered. What does the insecticide do to the young bee larvae? To the queen and her eggs? To the drones? What do non-lethal doses do to bees? Does it mess up their brains so they don't forage as well or impact their ability to find their way home? Does it make them more susceptible to disease?

Trying to address the problems of chronic neonicotinoid exposure and the long-term effects on hive health (and so all stages of bee life) is part of a new set of guidelines developed by the EPA, the Health Canada Pest Management Regulatory Agency, and the California Department of Pesticide Regulation. These tests are a whole new world for those who assess pesticides and try to keep bees safe. Designing and implementing tests that get at those sublethal effects are expensive, complicated, and difficult to interpret, and even more testing is on the way.

The New Brunswick bees were definitely taking a hit, which impacted blueberry yields and profits. What were the growers supposed to do? They were already suing, but the whole process was taking a while. What other options did they have?

They tried honey bees. Unlike today, in the early 1970s when Bridges Brothers was first having problems, hardly anyone used honey bees in the blueberry fields. To put honey bees on all 8000 of their acres would have cost Bridges Brothers nearly $400,000. Too much, Peter Kevan reported. I wonder if almost any price would have felt like too much. One year you have plenty of bees pollinating your blueberries for free and the next year you need to pay to get your blueberries pollinated? No one would want that. They'd want to turn back the clock instead.

I asked Peter Kevan, with forty-odd years to power his hindsight, if he thought the move to start using honey bees to pollinate blueberries was due to increasing pesticide use. He put it down to a variety of management practices that were changing around the same time. He said that some growers started using an herbicide called Velpar that killed most everything growing in the blueberry fields except the blueberries themselves. It seemed that the lack of alternate forage was impacting the bees. A study found that wild bees were having to travel further from their nests for food when the blueberries weren't in bloom, which stressed the bees. Also, Kevan said, the practice of burning the field every other year began to be replaced by cutting with a brush hog, leaving an area that was much less hospitable to the many ground-nesting bees.

I see this when Frank Drummond and I stop by a farm where a field had been burned the previous year. Frank says that last year thousands of bees had nested in the burnt ground. This year the growers mowed the other half, and when we go looking for nests, in both the old burn and the newly mown area, we find few. The bees had absconded. How far they went or whether they came back to this field to forage and pollinate we don't know.

Forty-two years

In 1976 Bridges Brothers won their case, but not for anything approaching the amount they'd asked for—they received $58,599. Nevertheless, they perceived it as a victory, particularly since fenitrothion, while still sprayed on forests, was no longer sprayed near blueberry fields. Another insecticide was used instead.

Controversy around fenitrothion use continued, though. Furor arose over a suspected link between the sprayings and Reye's Syndrome, a condition that causes swelling in the liver and brain, in local children. Throughout the 1970s studies in fenitrothion-sprayed forest areas saw diminished pollinators and lower seed set in at least some of the woodland plants. Despite all this, the aerial assaults continued but steadily diminished. By 1993 only 222,000 acres were sprayed, down from a maximum of more than 9 million. The decline was due more to diminishing budworm numbers than to a change in policy. Nevertheless, the fenitrothion was gradually replaced by other pesticides, mostly the biological agent Bt.

By 1994 the budworms numbers had declined enough that the spraying was suspended, at least for that year. That February, *The Hamilton Spectator* reported, "For the first time in the lives of many New Brunswickers, the only sound in the air over the evergreen forests this summer will be the drone of flies."

New Brunswick was the last province in Canada to use aerial spraying of fenitrothion. In 1995 Agriculture Canada, the federal department that licenses insecticides, said "enough." Four more years and then all aerial spraying of fenitrothion would be banned.

This particular outbreak of spruce budworm had lasted for forty-two years.

In July 2013 what looked like a freak summer snowstorm could be seen on Doppler weather radar in Quebec. It was actually a massive swarm of moths, the adults of the voracious spruce budworm. The Canadian government estimated one trillion of the moths had crossed the St. Lawrence River into New Brunswick. They were back.

Unexpected results

A big question I had going into this book was whether native bees can supply all the pollination a grower needs. For wild blueberries, the answer is—it depends. The reasons why the bees can't do the job in some lowbush blueberry fields do not usually seem to be about insecticides, despite what happened in New Brunswick.

According to Frank Drummond, the average number of insecticidal sprays on a Maine lowbush blueberry field is less than one per year (0.65),

On-the-job training for bumble bees

Bumble bees face a learning curve when they go out foraging. They need to figure out which flowers will provide the most nectar and/or pollen, and then they have to learn how to get it out. Researchers watched young *Bombus vagans* when they were out foraging. It took three to seven foraging trips for the young bees to figure out which plant had the best food rewards. When those flowers started drying up, the bumble bees clued in and went off to find the new best source. Sometimes they might pick the wrong plant because they hadn't figured out how to get to the floral rewards. For simple flowers like goldenrod, it doesn't take a bee long to discover how to get the goodies, but for some of the zygomorphic (snapdragon-like) flowers, it might take dozens of visits to become truly proficient.

with none during bloom. For comparison, highbush blueberries, the kind we commonly get at grocery stores, average three to six sprayings per year. Frank sent me this information and then added, "This is not to say that insecticides *do not* result in bee deaths in blueberry fields, but in a four-year study involving 40 fields we were not able to detect a negative impact on bee abundance or species richness in lowbush blueberry in Maine." I'm pretty sure this is science-speak for we can't prove a negative, but it sure doesn't seem like insecticides are regularly killing off bees in Maine's lowbush blueberry fields.

We know that other management practices in the blueberry fields do make life harder for bees. Large fields that are far from wild areas and their nest and flower resources, mowing rather than burning, clearing fields of flowering weeds that provide forage when the blueberries aren't in bloom, all these practices have adverse impacts on Maine's wild bees and their ability to pollinate the blueberries.

Some growers manage fine with the wild bee lottery, and a lottery it is. The quantity of bees can fluctuate up to ten fold between one year and the next. According to Frank, the organic growers like Theresa handle the poor years that come with these fluctuations pretty well. They have fairly low input costs, get higher prices for their berries (around $6 per pound for fresh berries versus about $2 for conventional), and often have value-added products like teas or jam to help buffer bad years. The super-high-input growers who bring in enormous numbers of honey bees, irrigate, and use herbicides and fertilizers are also shielded from wild bee fluctuations. However, for smaller scale, conventional growers a pollinator decline of even twenty-five percent can be a problem, and so these growers need supplemental bees to try and ensure good pollination.

Wild blueberries and local bees evolved together. Because of the way the fields are managed or where they are located, some growers can't make a profit without bringing in managed bees. If this is true for a wild native plant, what role do wild bees play on farms with more conventional crops? What role *can* they play? I went to California, where over forty percent of the fruits and vegetables Americans eat are grown, to learn more.

Cinderella *Ceratina* and Bees Down on the Farm

GORDON FRANKIE SITS in a classroom in the Carmel Valley of California, a tasty libation to hand, pinning bees. Pinning a bee involves sticking a straight pin through the thorax of a dead bee so that you can easily look at the bee from all sides when trying to identify it. Gordon has been studying bees and their interactions with plants for the last fifty years or so, and he's pinned a lot of bees. Nevertheless, to me, with his tanned skin, thin tidy mustache, colorful shirts, and baseball caps, Gordon looks more like a charter boat captain than an entomologist.

Some bees are killed in the name of science. People may get upset at the idea of killing bees to study them, but there are a lot of bees out there, so the losses are proportionally small and the goals worthy. Plus, some research simply can't be done without killing the bees. Field identification of live bees is difficult for most people, and even the best bee experts can't identify some small bees on the wing. To identify certain species, you may need to see their mouthparts, which are intricate, diverse, and really small, or the genitalia of the males, which they keep tucked safely away and out of traffic when not in use and have to be teased out. This is probably not something you want to try on a live bee, even if it can't sting.

Before a bee can be pinned and identified, you must first catch it. Bees are caught either in a net and then put in a vial with a killing solution (aptly called a kill jar) or they are captured in containers of soapy water (often called bee bowls) that they fly into, where they drown. The soapy

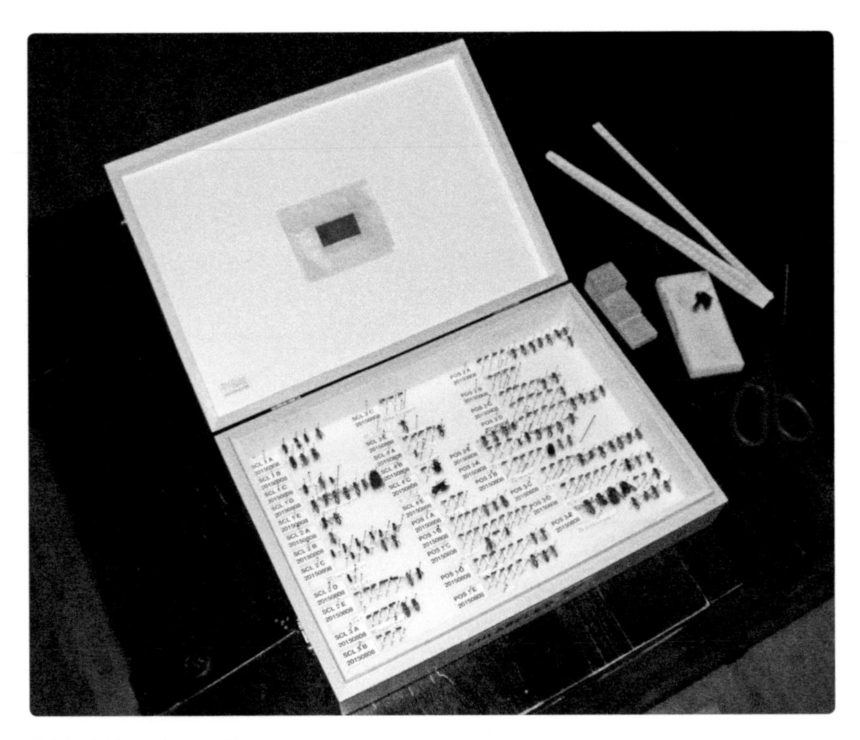

A Schmitt box of pinned bees.

water bees have to be processed before they get pinned: a rinse and a blow dry. The method I learned at Gordon's bee class involves putting the bees in a tea strainer (hopefully one dedicated to lab use) that functions as a tiny bee tumble dryer as you blow hot air from a hair dryer at the strainer. The purpose of the rinse and blow dry is to fluff up the bee's hair. You can see the colors better, it's easier to move the hair to look for markings, and, well, the bees just look better. I know they're dead and the last part shouldn't matter, but I've acted as a mortician for quite a few bees at this point, and I don't want them to be preserved forevermore in the midst of a bad hair day. So I coif dead bees. My children find me creepy.

After the bee is dry and as beautiful as you can make it, you carefully pin the bee to ensure it is absolutely level. On that day in the Carmel Valley, I was just learning to pin bees and some of my bees weren't quite level enough, so Frankie took them away to fix. I thought they looked pretty good, but clearly they were a bit off, maybe down by the head or listing a little to one side.

Once the bee is pinned, you might need to blow out its wings. Hold the pin up to your lips with the bee's bottom facing you and blow. This moves the four wings away from the body and spreads and separates them. Why? If the wings are wrinkled or folded over the body, important markings on either the wings or the body may be hidden. Other people in other labs have variations on how they do things, but the goal is to end up with a bee on which you can see everything that you need to in order to be able to identify it.

After the bees are pinned, you add three tiny labels. Each one is about the size of a pinky fingernail and has to be carefully cut out—each and every one. Then you stick the bees into a specimen box, which is often some version of a Schmitt box, a specialized wooden box that's maybe a bit smaller than a backgammon board, with a tightly fitting lid and a bottom coated in a material you can stick a pin into. (They run about $40 each.) So you spend an unbelievable amount of time collecting, coiffing, pinning, and labeling bees. You stake them out in their expensive boxes, and it's finally, finally time to identify them so you can actually do some science. Welcome to Gordon Frankie's world.

One day while I'm sitting in Gordon's lab at the University of California Berkeley, I start counting Schmitt boxes: boxes of bees on shelves, boxes on tables, boxes on desks. I count 157 boxes, each containing at least 100 bees. More than 15,000 bees—and those are just boxes from the last five years or so. I marvel at the number and not just because of the labor involved. During a bee survey, you only catch a small percentage of the bees that are out there. If this many got caught, imagine how many must be out there, flying about unnoticed.

Gordon Frankie is not a taxonomist like Robbin Thorp, nor is he someone who studies one kind of bee—not that he doesn't have his favorites. *Megachile fidelis*, Miss July on the Urban Bee Lab's pin-up calendar, is his favorite local bee. And his license plate reads "Centris," which is a big, showy bee that's uncommon in the United States. Gordon is a bee guy, but not just a bee guy. He's a where the bee meets the plant guy.

In recent years, a lot of his work has focused on bees in urban areas of California, and much of that information has been distilled on the Urban Bee Lab's website as a list of plants by season and the bees that like them. In 2009 Gordon decided to take all that he's learned about which plants California bees like best out onto the farms. It's a ten-year project aimed at learning what effect these plants can have on the farm. Does pollination of the crop plants increase? Does the farmer's bottom line? What do farmers need to know in order to be willing to devote some land, time, and dollars to pollinator plants? To try and figure this out, Gordon and his team conduct bee surveys, put in plants, water those plants when the farmers forget, educate farmers about the what and why of the research, and track the endless data. I've been curious about the role of wild bees on farms, and Gordon kindly offered to take me out to some of the ones where he's doing research to learn more. The answers I got weren't what I expected.

Brentwood, California: golf carts and fruit trees

The hour-long drive from Berkeley to Brentwood via State Route 4 skirts the southern edge of the California Delta and mostly seems to occur between walls. Periodically, the walls open up to views of housing tracts and commercial areas. I know it's not all Targets and automotive shops along this stretch of road, but that's what it seems like at sixty miles an hour. Not that long ago, it would have looked different—farmland planted in corn and stone fruits.

Brentwood itself has been a farming town since the gold rush days in the nineteenth century, but the farmers there jumped almost immediately into large-scale, commercial agriculture. At a time when the norm was to have small farms growing a variety of crops, Brentwood had large farms growing a single crop to provide for the miners. Today, farms are being replaced by houses and strip malls. Population increases and extraordinary housing prices in the Bay Area have driven people to the east. In 1990,

Megachile fidelis is Gordon Frankie's favorite California bee. This female is visiting *Cosmos bipinnatus* in Richmond, California.

7500 people lived in Brentwood; by 2005, it had ballooned to more than 50,000. Brentwood's history makes it a strange mix, with orchards lapping up against subdivision walls not far from the quaint old-fashioned downtown. Oddly, the reduction in farms and increase in housing may be good for bees.

Gordon and his project manager, Sara Leon Guerrero, work with eight growers in Brentwood, one of whom Gordon is taking me to meet, Farmer Al Courchesne of Frog Hollow Farm. Gordon and Sara work on various kinds of farms in Brentwood, and Frog Hollow is one of the certified organic ones. Gordon and I drive up to the packing sheds, and soon Farmer Al rolls up in his golf cart. Gordon says Al is always in his golf cart. Farmer Al is in his seventies but looks younger. He wears dark sunglasses, has a head full of silver hair, and flashes a movie star smile. Put him in a suit and he could be in an advertisement for watches or scotch, but apparently he

This showy *Centris rhodopus* female is visiting smoke bush.

never wears a suit. He wears overalls—all the time. I've been told that for more formal events he wears dress overalls.

Frog Hollow is about fruit, mostly stone fruit. Farmer Al started the farm in 1975 as a thirteen-acre U-pick place. When I visited, he had 215 acres (70 just acquired). The Frog Hollow website lists four kinds of cherries, half a dozen apricots and apriums, the same number of plums and pluots, seven kinds of pears, and eight kinds of nectarines. Peaches are king here, with fifteen varieties; it's the fruit Farmer Al started with. All these fruit trees benefit from bee pollination, but they can produce at least some fruit without it.

The need for an animal to perform pollination duties varies among plants. Some, like corn and wheat, don't need a pollinator at all; they are pollinated by wind. Some plants may be able to make fruit and seeds without pollinators, but with them the plants may produce more fruit or more seeds or larger fruit or, oddly, fruit that lasts longer. Other plants flat out need an animal, sometimes a very specific one, to procreate at all.

A study in 2007 found that of the top 115 global food crops, 87 needed animal pollinators, but to varying degrees. Plants were divided into groups based on how badly they needed those pollinators. For 13 crops, pollinators were considered essential; for most varieties of that plant, production is reduced by ninety percent or more with no pollinators. The next category, high, saw a forty to ninety percent reduction. This second group includes most of the Frog Hollow fruits. A plant in this category might be able to reproduce in the wild without pollinators but good luck trying to make a business growing them without animals to carry that pollen around.

The bees doing a good chunk of this pollen transport aren't always the big charismatic ones like bumble bees or BOBs. Open one of Gordon's Schmitt boxes and there will undoubtedly be rank after rank of small to downright tiny bees. Selling the wonders of a bumble bees or a glorious green *Agapostemon* is fairly easy. From a beauty point of view, a lot of these little bees—lack. That doesn't mean they don't have an interesting story to tell.

Ceratina bees, unlikely pollinators

Although Gordon Frankie and crew are studying all the fruits out on their Brentwood farms, some are getting looked at more intensively than others. Gordon tells me that they've done a huge amount of background work on which bees are visiting the berries. He says that Sara has "almost two hundred counts already. . . . In addition to honey bees that are visiting were all these *Ceratina* . . . and these little *Ceratina* are just full of pollen, so that's one of the focal points for our research right now. We were told years ago that these bees were good, but no one ever went out and did the data collection."

Little *Ceratina* a pollination powerhouse? That's unexpected.

Members of the genus *Ceratina* are carpenter bees—bees that chew wood. Carpenter bees in the United States come in two sizes: big and small. Writing in 1928, Phil Rau had some problems with placing the small carpenter bee into the same category with the large carpenter bee, based on the power of their mandibular carpentry tools. He writes enthusiastically of the large carpenter bee carving out tunnels in solid wood with "joyful vigor." The small carpenter bees, he says, don't even really deserve to be called carpenter bees since they can only burrow into the soft pith of plants like elderberry or sumac. Even then they can't get in unless there's already a hole or a break in the stem.

Poor little *Ceratina*, to be so scorned. What else did Rau expect? These bees are small, running about the size of a rice grain, and a long skinny rice grain at that. They'd probably have a hard time flying weighed down by mouthparts that could chew through proper wood. Although given some of the excessive antennae, long tongues, and massive loads (pollen, mud, leaves) bees are able to carry, perhaps I am doing *Ceratina* a disservice.

Most species of *Ceratina* are not showy bees. They have little hair and usually come in basic black. Many could be described as understated, although they do tend to have a nice shine and some actually gleam a metallic blue, green, or bronze. To me a lot of *Ceratina* bees look like their close relatives, the ants, only with wings.

Ceratina bees are not alone in being small, blackish, and understated. *Halictus* and *Lasioglossum*, both sweat bees, and *Hylaeus*, yellow-faced bees, are three other common ones. I was out in the field with an old bee biologist once, and a bunch of little bees were zipping about. He said they were halictids (sweat bees). They were tiny. I wondered how he knew that they were halictids and not, say, *Ceratina* or *Hylaeus*. So I asked him. His response was something along the lines of "they have a certain gestalt." Gestalt? Well, pish, that's not going to help me learn how to identify them.

I don't know if *Ceratina* bees have a gestalt, but they do have a distinctive look that is pretty easy to recognize with a little practice. It's all in the abdomen, at least if you catch one and apply some magnification. Sam Droege at the U.S. Geological Survey Bee Lab, many of whose bee photographs are in this book, says *Ceratina* bees have "an abdomen ribbed like a plastic water bottle." If you ever get a chance to lay a bunch of little black bees out, start by looking at their abdomens and you'll begin to see the difference.

Despite all the *Ceratina* bees and the fifty or so other species that Gordon and Sara have found at Frog Hollow Farm, Farmer Al still brings in honey bees to pollinate. Gordon isn't even proposing that his plantings will allow Farmer Al or any of the other farmers to be able to do away with honey bees. Gordon says he doesn't have enough land to work with for that. He sees the bees his plants support as supplements. Nevertheless, I've wondered since I decided to write this book if the honey bees all upped and died, could the native bees take over pollinating our farms? They certainly did once upon a time.

A short history of American agriculture in the twentieth century

America, 1900—Around half the U.S. population lived on farms and about forty-one percent of the labor force worked there. The main purpose of the farm was to provide what the family needed as well as a crop or two that was good for sale so they could buy the things the farm didn't produce. The power to wrest crops from the land came from muscles, of both people and animals. Over 21 million work animals, primarily horses and mules, provided most of the power, pulling plows in the field and carts into town, but people labored as well: cutting wood, hoeing, tossing hay, milking the cows, slaughtering the hogs, pumping up water. The farms were mostly small, averaging around 100 acres. All those animals needed food, and so farmers had pastures and grew hay and oats. Fuel to cook with and heat the house came from the wood lot. The diverse land use meant plenty of nest sites and floral resources for wild bees. Plus, many farms also had a honey bee hive or two, whose primary job was to provide honey and beeswax. People ate, so clearly the pollination was adequate for their needs.

America, 1950—By the end of World War II, life had changed. Many farms had electricity. Tractors had replaced work animals (and humans) for much of the power needs on the farm. By 1945 the number of work animals had been cut almost in half, replaced by 2.4 million tractors. Because tractors don't need oats or hay or a pasture, that land could be put

Ceratina acantha, a small carpenter bee, visiting *Mimulus aurantiacus* in El Cerrito, California. ▲

A rare photo of the underside of a bee. This *Ceratina dupla* was found in New York State. ◄

to other uses. A variety of implements could be attached to the tractors so they could do many jobs. Milking machines had come along. At the same time a vast infrastructure of reliable roads had been built. Refrigeration allowed produce to travel further from its source. Farm products could be transported and sold, the money used to buy all sorts of goods. Synthetic fertilizers and pesticides had come along, as had new seeds like hybrid corn. The increased yields available with added nutrients, new seeds, and fewer pests combined with tractors allowed the same amount of land to produce more with fewer people. By 1950 the primary purpose of a farm wasn't to feed the farm family but to provide commodities for sale.

So, most of the massive transition in agriculture and Americans' way of life happened in just the first fifty years of the twentieth century, but agriculture continued to evolve. Farm output over the course of the twentieth century quadrupled while the percentage of the population working the land plummeted to less than two percent thanks to all the changes in technology. Between 1900 and 2002 the number of U.S. farms decreased by sixty-three percent while the average size increased by sixty-seven percent.

What's grown on a farm has changed as well. Individual farms grow fewer crops, often only one. One gigantic field of all the same crop is going to be easier to farm than the same land area broken up into a patchwork of different plants. Plus, several of these monocrops, including corn and soybeans, have financial safety nets provided by the federal government, protecting farmers from a catastrophic year.

These farming changes weren't good for bees. From 1929 to 1963 Everett Oertel, an apiculturist for the U.S. Department of Agriculture, tracked the weight of honey bee colonies in Louisiana throughout the year. During those years, honey bee colonies went from an average weight gain of seven pounds between September and November to an average weight loss of twenty-four pounds in the same period. He attributed the massive decline to the use of herbicides and changes in pasture care that reduced the fall honey flow from goldenrod. He also thought that growing more crops that weren't as attractive to bees and increased urbanization played a role in the decrease. The honey bees were going into winter with much smaller reserves. Presumably some of the changes that impacted honey bees also made life hard on the wild bees, who lost not only the flowers but weedy-edged nesting sites as well.

Large fields, easy tillage by tractors, herbicides, Roundup-ready plants that allow for yet more herbicides, monocrops, all make life easier for farmers but harder for all kinds of bees. Fortunately, for the diversity of our diet and the potential happiness of our bees, some farmers, like Farmer Al, choose a different route.

Back at Frog Hollow

After climbing out of his golf cart, Farmer Al takes us up to a big office space above one of the packing sheds. There, I hear about one man's way of running a successful farm and what it takes to get a piece of fruit to market. Little of our talk is about pollination. I hear about Al's compost project and, of course, the weather. When we tour the farm later Al points out an area and tells me that not long before, the ground was covered in gold. A few days of heat, some rain and wind, and the fruit just started dropping. Fifty thousand dollars lying on the ground.

Diversification seems to be a key to Al's success, both in what he grows and how he sells it. Al grows dozens of cultivars of a wide variety of fruits and seems to be constantly looking for new ways to sell it. Frog Hollow fruit is sold at a farmer's market and through community supported agriculture. It's shipped to grocery stores or to your door; I like the sound of the monthly "Cornucopia of California." Frog Hollow runs a cafe and has jams, granola, cookies, and other baked goods for sale. Farmer Al likes projects, and this liking expands his options.

We do, eventually, manage to talk about bees and the pollination project. Al says his favorite bee is the carpenter bee, an understandable fondness. The valley carpenter bee (*Xylocopa varipuncta*) is nothing like the little *Ceratina* carpenter bees dashing about Al's farm. *Xylocopa varipuncta* is a whopper of a bee, and the male is a fuzzy sweetheart, who, being male, doesn't sting. Gordon and his lab crew call them teddy bear bees and use them as a one-bee petting zoo if they have one handy when they're giving talks.

When I ask whether Farmer Al thinks Gordon's plantings and the wild bees are making a difference, he says he doesn't know, but "the numbers keep going up on crops that need pollinators. Pluots, last year was a record year but this year will beat it by a long shot." Suggestive, but not a totally satisfying answer. Pollination is always more complicated than I want it to be.

Farmer Al offers to take me on a tour, and Gordon and I hop in the golf cart. We bump down a row between peaches. Al has 35,000 trees, all of which need pruning, fertilizing, thinning, and picking. "For every single piece of fruit you see, we removed at least two," he says and points to the ground where I see the litter of tiny fruits. In the trees, the yellow and red winners of the thinning lottery glow. For each acre it costs about $1000 to prune, another $1000 to thin, and another $1000 to pick. That's 145 acres, in the process of expanding to 210 acres. "Think about it," he said. "Millions of fruits." And most of those fruits—those that were kept and those that were pinched off—are the result of bee visitation.

As we drive down the rows, I realize this orchard is different from many that we passed on the way here. Those had oceans of clean, bare dirt. Al's rows are full of grass and other plants people would just call weeds. Farmer Al says other growers come by and are appalled. How can he stand the mess? It doesn't bother him, and now he's learning that a lot of those plants are good for the bees. Even as we putter along, Gordon points out weeds that are good bee plants. He extols an ugly-looking yellow composite called *Picris* and points out another, *Verbesina*, saying, "Don't cut that down."

Clearly, the farm is a good place for bees. There are fifty species here, after all, and all those zillions of *Ceratina* bees. The bees have lots of food options—dozens of kinds of fruit flowers, weeds, Gordon and Sara's plantings—and the farm is organic. It seems like this would be a farm capable of providing Farmer Al and family with a decent living without honey bees. But chances are, it can't.

Thievery

This male *Xylocopa varipuncta*, the valley carpenter bee, from California is robbing nectar. He's too big to climb up inside the flower, and his tongue is too short to reach up from the bottom, so he slits a hole and steals. His back door route avoids contact with the stigma and anthers, and so the plant loses out on a pollination opportunity. It's even more unfair to the flower than it sounds, because sometimes other bees come along and say "Woohoo, easy access" and use the hole too.

A *Xylocopa varipuncta* male, the teddy bear bee.

A place to live and food to eat

To have a wild bee visit a crop flower, the bee has to find an appropriate nest site within its flight range of that flower. Some bees can fly quite a ways, a couple of miles if they have to, but of course they'd rather find a grocery store closer to home. Other bees can only fly a few hundred feet, which means nest site and floral grocery store *have* to be close together. You can plant good bee flowers all day long, but the bees must have a place to call home too: old stems, punky wood, sandy flat ground, well-drained slopes, what have you.

It's easier to focus on providing bees with flowers rather than nest sites. For one thing, most of us like flowers. Bare ground and old stems aren't so exciting. Plus, we see the bees using the flowers, which is gratifying. A farm or orchard may have loads of flowers, but often only for a few weeks, since many farms grow only one or two crops. The same thing can happen in a garden with a bonanza of bloom in summer but not much early or late in the season.

Bees need a variety of kinds of flowers available over the course of the year. If one kind of bee lives for a month but the nearby flowers are only available for two weeks, that's a problem. Plus, different kinds of bees come out at different times of the year and all of them need food. Variety is important too. Chances are that most bees (who aren't pollen specialists) do better with an array of flowers to choose from since all pollen is not created nutritionally equal. Also, an assortment of flowers needs to be available in each season because not every bee can access every flower's pollen and nectar. A bee's tongue may be too short to reach into a flower, or a bee may be too small to open the petals and get at the goodies. So, for a farm to get pollination from wild bees, it needs adequate amounts *and* a variety of both flowers and nest sites—all within the flight ranges of the bees. Suddenly, having ample wild bees to pollinate crops well enough to make money doesn't seem so straightforward.

Cinderella *Ceratina*

Consider the lifestyle and needs of *Ceratina*. These little bees are unlikely to travel far from nest to food, so in order for a farm to have *Ceratina* bees it needs nest areas near the crops. *Ceratina* bees nest in pithy stems, like raspberry and elderberry. A *Ceratina* female finds a dead stem, either broken or with a hole in it, and burrows in and starts chewing through the pith. After she gets a nice long burrow hollowed out, she starts

Bee tongues

To talk of bee tongues is to vastly oversimplify the complex apparatus that make up bee mouthparts. Bees have a glossa, which is the closest mouthpart they have to a true tongue. They also have labial palps that run next to the glossa and are used for tasting. These, together with other mouthparts, make up the bee's proboscis. Many of these parts are jointed so they can be folded up.

Some bees are called long-tongue bees and others short-tongue bees, but it has to do with the labial palps rather than the glossa length. This leads to the confusing truth that some short-tongue bees have a long glossa and so could be considered long-tongued, short-tongue bees. I won't take that thought any further, but a cool thing about bee mouthparts is that the ones with long tongues fold them away under their body when they aren't in use. When needed, they can unfold the pieces and join them together to make a straw for sucking up nectar. Short-tongue bees don't make the same kind of straw and may lap rather than suck up nectar.

Andrena nivalis, with a short little tongue, from Pictured Rocks National Lakeshore, Michigan. Bee mouthparts vary widely depending on how they're used.

This female *Anthophora affabilis* with long mouthparts was found in Badlands National Park in South Dakota. Clearly, this bee and *Andrena nivalis* are likely to feed off of different flowers.

This bronze and black beauty, *Ceratina mikmaqi*, was found in Maryland and is only as big as a grain of rice.

provisioning. She forages, getting enough pollen and nectar for one egg. She lays the egg and caps the cell with bits of pith scraped from the side walls of the stem. Then she starts on her next cell. This one burrow in this one stem is the only place that she will lay eggs. It's also home. She stays there and guards against predators and parasites. This is where the *Ceratina* story get interesting.

Bee lifestyles range along a spectrum from the solitary, like BOBs, to the highly social, like honey bees. The highly social bees have multiple generations living together, dividing up the work, and caring for the babes. Social bees evolved from solitary ancestors, but most bees don't evolve in that direction. Until fairly recently, *Ceratina* species were primarily viewed as standard solitary bees, laying in all the provisions the baby bee

Regretfully, this glorious small carpenter bee, *Ceratina smaragdula*, has invaded the Hawaiian Islands. ▲

Gleaming bronze with a hint of green, this *Ceratina* was found in the Bay Area of California. ◄

will need to grow from egg to adult and then leaving the babe to get on with it. Spectrums, by definition, have a range, and it turns out that *Ceratina*, depending on the species, falls into several of the in-between places on the bee sociality spectrum.

To become social, a bee has to be genetically capable of a few things. If a type of bee doesn't have the prerequisites, its descendants will never become social. First, the bee has to have a fairly long adult life, because caring for the young is part of being a social bee. Second, the bee has to be loyal to the nest. If the bee lays eggs all over the place and never goes back to those places, clearly no childcare will happen. Lastly, the bee needs a tolerance for being in close company with other bees. At least some *Ceratina* species have these prerequisites and have taken a step or two along the spectrum away from a truly solitary lifestyle.

Ceratina calcarata lives in eastern North America. The mother bee lays all her eggs in one stem and stays there to look after them. She goes in every evening and checks on the kids, grooming them and removing parasites or fungus. Sandra Rehan at the University of New Hampshire has spent a lot of time with *Ceratina* bees. She says that the *Ceratina calcarata* she's studied in Canada have all gone through metamorphosis by around August but remain in their birth stem as adults, along with mom. The newly hatched adults don't go out for food or to have sex or to start their own nest; that all happens the next spring, assuming they survive the winter. To survive the winter, they need to keep their fat stores up. So mom goes out and forages for them, but she doesn't do this job alone.

When the mama bee lays her eggs, her eldest daughter gets a smaller pollen wad made of less nutritious food than the others. As a result, this firstborn daughter will be born small. Sandra refers to her as "dwarf eldest daughter." Once born, this daughter is pushed out of the nest by mom to go forage for her siblings. Because of her small size and the work that she does, the dwarf eldest daughter doesn't survive the winter and so never has a chance to have babes of her own. Some have referred to her as Cinderella, but clearly she is Cinderella without a fairy godmother and no chance to go to the ball.

Not all *Ceratina* bees live this way, and only a few of the 350 or so *Ceratina* species in the world have been seriously studied. No one knows if any of the *Ceratina* bees out collecting late in the season at Frog Hollow are little Cinderellas, but we do know that however much pollen they and the other native bees gather at Frog Hollow, they likely don't meet the pollination needs for a successful farm because of where Frog Hollow is located.

Bee-friendly farms: it's not necessarily about the farm

Claire Kremen, Neal Williams, and Robbin Thorp investigated whether wild bees could get the pollination job done in California watermelon fields. They chose watermelons because they are highly dependent on pollinators (one of the thirteen on that "essential" list). They published their research in 2002, after varroa mites had taken hold of the honey bees but before CCD hit. At that point, for all anyone knew, wild bees might already be doing a lot of crop pollination—or they might not.

They looked at how much pollination wild bees were doing on four kinds of watermelon farms. The farms were either organic or conventional (that is, used synthetic fertilizers and pesticides) and either near or far from a natural area. Being near a natural area meant that within a one-kilometer radius of the farm thirty percent or more of the area was natural. If a farm was far from a natural area, then less than one percent of the area within that circle was natural.

Kremen and her colleagues found that wild bees could meet the pollination needs for watermelons (provide 500 to 1000 pollen grains per flower) for the organic farms near natural areas, but that was it. Organic alone wasn't enough. Being close to wild areas seemed to play a more important role than the organic status. They found that agricultural intensification—more agricultural fields, less wild area—meant fewer bees overall and fewer species as well. Both matter when it comes to pollination. Needing enough bees around to pollinate flowers is obvious, but the benefit of different kinds of bees is less so and can show up in a variety of ways. Some bees get

Females of the squash bee *Peponapis pruinosa* often like to nest in the ground below squash plants. The males sleep in the flowers.

pollen to different parts of the stigma, making for better fruit. Some years the timing may be off for a particular type of bee—it may not be out when the crop is blooming—or some other factor just makes it a bad year for one particular kind of bee, so having other bees that are available is key.

Gordon's research over the years has shown that there is another potential reservoir for bees: residential neighborhoods. The farms Gordon works with in Brentwood are scattered around. One farm has a greater diversity of bees than that of Frog Hollow, despite being a conventional farm. Gordon attributes this to the nearby residential areas with their gardens and all the flowers that they contain.

A look at Google maps shows that, except for a small cemetery, Frog Hollow is surrounded by farmland. The monochromatic rectangles of brown and green suggest large fields of one crop—poor reservoirs for bees.

A *Halictus tripartitus* male in Berkeley, California. These bees are tiny but sometimes stupendously numerous, and Kremen's team found them out in the watermelon patches.

Al's farm may not be able to supply enough bees to pollinate all his plants adequately without the help of honey bees, but his weedy rows and Gordon and Sara's plantings may help his bottom line and certainly provide an oasis for bees in the area.

The power of wild bees

Either Gordon or Sara goes out to the Brentwood farms at least once a week during the growing season. They water, they plant, they count. They collect bees to fill more Schmitt boxes, and they share the information they glean with the farmers, hoping some of that information will lead to changes that are good for bees. One thing they found was mind-boggling to me.

"You do frequency counts," Gordon says, "and you get the average number of bees on a tree—honey bees—is somewhere between three to seven bees. . . . At any given time on any given tree that's all you get." And

Gordon takes that information to the farmers. "Okay, you only get one native bee in the tree and you've got five honey bees but, by God, probably that one [native] bee is doing more pollination than the five honey bees."

That's it? A handful of bees at any one moment provide enough fruit to fill the stores and keep a farmer in business?

Research confirms the power of wild bees. Usually scientific research papers have a few to maybe half a dozen authors. So, when I saw a paper with a whopping fifty authors, it got my attention. One thing that number of authors tells you is that that there was a remarkable amount of agreement on the research involved. This particular study reviewed the effect of wild pollinators on crop yields in forty-one crop systems all over the world. The authors found that fruit set increased in every one of those systems when they were visited by wild pollinators, but only fourteen percent of the systems showed significant increases in fruit set with honey bee visits. The authors had predicted that wild bees would increase fruit set only when there weren't many honey bees around, but they found that wild bees improve fruit set even when there were lots of honey bees.

Part of the power of wild bees is their diversity. They fly at different times of the day. Some will fly in bad weather. They forage differently and contact the stigma differently. They carry pollen differently. If only one pollinator, like a honey bee, is available, it makes sense that it would not be as effective as a bunch of different kinds of bees.

A study done in apple orchards in New York State found that a couple of things affected pollination. Not surprisingly, greater species richness mattered. The researchers also found that having an assortment of *types* of bees visiting the apples improved pollination. One can take several attributes of bees—for this study it was size, sociality, and nesting site—and group them into a whole slew of types. Some bees, like bumble bees, are big social cavity nesters, Others, like some species of *Andrena* and *Lasioglossum*, are small solitary ground nesters. The researchers found that having a variety of types was even more important than having lots of species. Seed set nearly tripled in orchards when the number of types of bees visiting increased from one to more than four. What's more, they found that a lot of the pollination happening in the apple orchards that had been assumed to be from managed honey bees was likely being done by the home team, those wild bees living their lives on the surrounding land, hidden away in stems, holes, and underground nests.

Native bees *do* make a difference on the farm, even if sometimes they can't do it all.

Gordon's plantings at Frog Hollow are small, tucked in at the end of rows, or maybe filling in an acre or two after a bunch of unsatisfactory trees get pulled out and the new ones haven't gone in yet. Those smatterings of habitat surrounded by a sea of farm fields aren't sufficient to support enough bees to replace the honey bees, but they may be helping Farmer Al's bottom line. They are also building little bits of habitat in farming country and perhaps providing the beginnings of stepping stones between wild areas. Clearly, those wild areas make a big difference in providing pollination services for farms lucky enough to be close to them.

What about all those wild areas? How well are bees holding up there? I've spent the first part of this book focusing on bees in agriculture. It's time to move on—out to the bees in the wild.

Life, Death, and Thievery in the Dark

THE LAB FEELS like someone's porch—all wood and windows with a screen door—but a porch that has been overtaken by a group of mad bee enthusiasts. Every horizontal surface is covered in bee paraphernalia: boxes of bees, a little fish net for rinsing bees, reference books, boxes of pins, pinning boards, a gallon jug of detergent for bee bowls, forceps, notebooks, someone's bee net, microscopes big and small, and a hot air popcorn popper. Of all the bee paraphernalia in the room, the popcorn popper is by far the coolest.

All this gear belongs to people who have come to the American Museum of Natural History's Southwestern Research Station in southern Arizona for the Bee Biodiversity Initiative gathering. The initiative, according to its website, is "a collaboration among bee biologists from all intellectual arenas to better understand bee diversity and ecology in understudied areas." One of those collaborators is Denny Johnson, a retired production engineer, and he's the one who has taken a hot air popcorn popper and MacGyvered it into the ultimate bee dryer.

Bees that are caught in bowls of soapy water (bee bowls) need to be rinsed and dried. Before seeing Denny's modified popcorn popper, I had thought that using a tea strainer and a hair dryer was the height of bee drying efficiency. Ha! Denny's popcorn popper holds more bees, and you can just turn it on and leave the bees to bounce themselves dry in gentle puffs of air while you go do other work. Or you can watch them while they bounce; it's mesmerizing. (Do not try this at home with a regular popcorn popper unless the goal is burnt and burst bees.)

I hadn't planned to be at this Bee Biodiversity Initiative meeting in May. I'd hoped to come to the research station in August instead to attend the Bee Course, nine days spent collecting bees and learning how to identify them. I realize that probably sounds nightmarish to most people, but I was looking forward to it. I didn't get in. Amazingly, more than sixty people had applied for the twenty-four spots available. Fortunately, Jerry Rozen, who started the Bee Course and still does a lot of the running of it, offered to let me come down in May to hang out with the Bee Biodiversity Initiative folks. It might not be the Bee Course, but it was still a group of bee people meeting in an area with some of the richest bee fauna in the country, and I was invited. I went.

The Southwestern Research Station

The research station is located in the southeastern corner of Arizona in the Chiricahua Mountains, one of forty or so sky islands scattered about the region where Mexico, Arizona, and New Mexico come together. The sky islands are aptly named. They form an archipelago, with each island surrounded by a sea of hot desert and scrub rather than water. Climbing up one of these islands takes you through climate zone after climate zone, ending, for the tallest ones, in cool conifer forest. Sky islands exist around the world, but these particular ones, the Madrean Sky Islands, are unusual because they are stepping stones between the Sierra Madre Occidental of Mexico and the Rocky Mountains in the United States, so in the Chiricahuas the edge of the tropics washes up against the cool-weather Rockies. Birders flock to the Chiricahuas to see tropical birds found nowhere else in the United States, and bee hunters swarm the desert scrub for its wealth of bees large and small.

This isn't my first trip to the Southwestern Research Station. I came last year for a few days at Jerry's invitation. That time, after the long drive from Tucson, I arrived close to dinnertime and met Jerry, his graduate student

The American Museum of Natural History's Southwestern Research Station near Portal, Arizona. ▲

This *Martinapis luteicornis* was found in southwestern Arizona. This species has been seen out gathering pollen before sunrise. Perhaps it wears such a fine fur coat to keep the lingering chill of the desert night at bay. ◄

Anna Holden, and Glenn Hall from the University of Florida. After dinner we went to the lab, because that's what you do after dinner at the research station. That day had been a good bee day for Jerry and his crew.

I still remember my astonishment when I sat down at a microscope and looked down the tube into a Petri dish filled with a jumble of bees. Black bees, gray bees, buff yellow pollen, a purple petal, tangled legs. Bees lying on their back with their legs crossed. One bee's tongue was out, looking as long as its body. Eventually, the jumble started to resolve into some recognizable bees. Megachilids with fat bunches of hair under their abdomens. Long-horned bees with such straight long antennae that they made me think of African antelopes. Why *does* a bee need such long antennae? Bees covered in pollen. The pile glowed with sparks of color from iridescent green *Agapostemon* and the shiny enameled yellow of *Perdita*. The bees were beautiful. It doesn't seem like they should have been, all dead and tangled, but they were. They also represent a small, small number of the huge diversity of bees that live in this part of the country.

Many people come to this corner of Arizona to chase bees on the wing. Not Jerry Rozen. He has devoted his life to studying life in the nests of solitary bees: eggs that hatch into pale still larvae that eat and wait for their moment in the sun. Sometimes, they never get that moment, because an interloper enters the nest and then there is a battle and death in the darkness. Jerry studies it all.

I asked Jerry once how many nests he'd found over the years. He didn't give me a straight answer. Instead, he picked up a copy of Michener, McGinley, and Danforth's *The Bee Genera of North and Central America*, opened it, and started pointing. I tried to keep up, and this is the best I can make out from my notes: eight types of *Calliopsis*, eight nests of *Nomadopsis*, some other species that I missed, a *Hypomacrotera*, several species of *Heterosaurus* (oops, no: there are no bees called *Heterosaurus*, plus that sounds like a dinosaur rather than a bee), some *Panurginus*, numerous species of *Perdita*. Then he turned the page. I got the idea—he's found a lot of nests.

I'd rarely thought about the life of solitary bees in the nest. Bee larvae are white maggoty-looking things. They don't seem to do anything other than eat. I considered them boring and ugly. Also, although I know they *are* bees, they don't *feel* like bees. That's the problem with complete metamorphosis. We think of voracious caterpillars as something completely unrelated to beautiful butterflies, and I've managed to do much the same with adult bees and their larvae. Regardless of my attitude, the reality is that most bees spend eighty-five to ninety percent of their lives in the nest. Jerry told me once that people think they know about bees because they know about the adults. They take what they know from that one life stage and "make these leaps." It's hard not to make those leaps, because most of us never see a bee that isn't an adult.

I'd assumed that my second trip to the Southwestern Research Station would be similar to the first: out with my net collecting bees flitting in the sun, hanging out with bee people, and learning more about the bees of this area. Instead I ended up heading into darkness, into the world of bee nests. More goes on there than I ever expected.

Hunting baby bees

Three people stand in a wide open landscape of yuccas, cacti, sundry dried plants, and dirt. The wind blows. No one speaks. A fourth person sits on a dirt seat carved into the side of a hole, wedging little slivers of dirt off the wall with a trowel, periodically pausing to blow loose dirt away with a special little hose. Work stops every now and then to shovel out the bits of dirt that fall to the bottom of the hole. Wind, silence, digging, waiting. When the digger sees evidence of a nest cell, he switches to a penknife and carefully, slowly, surgically removes infinitesimal bits of soil to uncover the tiny delicate baby bee larva.

With forceps, he holds it aloft for all to admire. We should have a balloon, "It's a larva!"

The larva doesn't squall at being removed from its cozy sleep space. It hangs there, white and unmoving. It gets placed in its bee bassinet, a

Teeny *Perdita*

The genus *Perdita* consists of about 700 named species; all are small to downright tiny bees. Most are pollen specialists of some sort, collecting from only one or a small group of flowers.

Behavior choices in bees can be perplexing. Two solitary species of *Perdita* that live in the same part of southeastern Arizona nest in the ground in the same area, take three trips to provision one nest cell—two for pollen, one for nectar—and appear to like pollen from the same plant. Their behaviors seem practically identical. Yet one of the bees, *Perdita difficilis*, takes a southern Mediterranean attitude to foraging. The females go out to collect between seven and ten o'clock in the morning and then take a siesta during the heat of the day. They head back out between five and seven thirty. The other bee, *Perdita luciae*, starts at eight but motors right on until about one in the afternoon before calling it a day.

The head of *Perdita luteola*.

special larvae-holding box, with little larva-sized divots into which the babe is curled. Jerry Rozen had someone at the American Museum of Natural History create the dish to his specs. The careful digging begins again. Hunting baby bees; Jerry has been doing this for sixty odd years.

Although he works in New York City, for most of those sixty years Jerry has come to this vast rugged corner of Arizona at least once a year to hunt bees. Often his goal isn't the adult bees that most people seek. Oh, he seeks them, but only so they can lead him home to the babes. On my first full day at the research station, three of us are preparing to go on a baby bee hunt with Jerry: me, Corey Smith (Jerry's assistant), and Joan Milam of the University of Massachusetts Amherst. The car is loaded with all the items needed for a day in the field: jars, nets, bee bowls, detergent, a field microscope, the weird dirt-blowing tube, the special larvae holding dishes, forceps, notebook, hand lens, digging implements, lunches, cameras.

The first difficult step in hunting baby bees is having some idea of where in this wild empty land a nest site might be. When looking for the nest of a specialist bee, you at least have a place to start: find the plant(s) the bee likes. We do this on another day, spending an entire morning searching for a specific bee that specializes on creosote bush (so named because its leaves smell like coal tar) in the hopes that we can track one back to its nest. We drive along a stretch of road looking for creosote in bloom, and when we find some we hop out and look for our bee. When we find no sign of it—which we never do—we climb back in the car and head to the next set of creosote bushes. The rest of us might start to get sidetracked by other bee possibilities, but not Jerry. This single-mindedness is probably part of why he's found so many nests over the years.

When (or if) an adult bee is sighted, the next step is finding the exact location of a nest. If the nesting needs for that bee are known and fairly specific—stems of yucca, perhaps—you can go to likely sites and hunt. If a bee likes flat sandy ground, that's going to be more of a problem here in the desert scrublands. So the other option is to try to follow a jigging bee as it flies from flower to home. If, after all of this, a nest is found and it's in the ground rather than an aboveground stem or hole, the careful digging through the layers of dirt begins. Sometimes the burrow is open and you can blow dry plaster of Paris down the hole and follow the treasure map of white powder, which will hopefully lead to the larval gold. Other times the burrows are closed off, so you just dig, oh so carefully, hoping to avoid cutting off random larval parts with trowel or penknife.

Fortunately, on this day Jerry had already dealt with the first two steps. He knows where a nest is. In August 2015, he had seen multitudes of the adults flying over a huge field of yellow daisy-like flowers (*Heterotheca*) and had marked a nesting area. We seek the offspring of those bees. The species we're after is *Hesperapis rhodocerata*, and Jerry says they are odd bees. Prior to 2015 the last time anyone had seen this species in the area was in 2010, and he and others had looked for the bees every single year since then. Where had they been? Jerry wonders if they can do the cicada thing and wait underground for years for conditions to be right. No one knows for sure, but this would fall into the unusual bee behavior category.

So, on my first nest-hunting day, we drive down from the trees and greenness of the research station out into the scrub, and we keep on driving. We pass a few cows, a sign for a ranch, a cinder cone, scrub, more scrub. Eventually, we stop in the middle of nowhere. Jerry gets out of the car, heading off purposefully through the dried grass, and walks right up to the nest site. At first I'm amazed he found it in all this sameness, but it turns out that once we got close to the spot he'd been counting telephone poles. The rest of us follow him to the nest site and dump our gear. Jerry grabs the shovel and starts digging a hole. He's eighty-eight. Eventually Corey gets the shovel away from him and continues digging the hole, carving out a comfy seat. Then the careful chiseling out of dirt clumps begins.

So there we all are. Three people stand, or squat, or sit on the ground, watching a fourth chipping away at the side of a hole. We spend at least an hour there, rarely speaking. The occasional find of a larva is surprisingly exciting, despite its inertness. After we've found several larvae, Jerry calls it a day. We head back to the lab early, Jerry carrying his plastic-covered dish of larval booty.

A *Hesperapis* female loaded with pollen on *Encelia farinosa* (brittlebush). She shows the flattened abdomen characteristic of the genus.

Digging for baby bees in Arizona—Jerry Rozen digs while Corey Smith and Joan Milam wait their turn.

Life in the lab

After a day of collecting, everyone gathers on the patio outside Jerry's room for a drink before dinner. I think Jerry instituted this cocktail hour during the Bee Course to make people stop with the bees for a bit and interact with each other. After dinner, it's off to the lab to process the day's bees and hopefully find something new and exciting.

Every night I'm at the station I too go to the lab. I work on my computer and peer at the bees that people are admiring. I cruise through the reference collection, drawers of labeled bees that can be compared to unknown bees to help identify them. The drawers have not only species I've never heard of but whole genera that are unfamiliar. There's a thrill to inspecting an unknown bee. What makes this one special?

I look for and find an adult of the species whose larvae we'd dug up, *Hesperapis rhodocerata*. It has the characteristic *Hesperapis* flattened abdomen plus short little antennae. *Perdita minima*, believed to be the tiniest bee in the United States, is there too, all shiny caramel with long curling antennae. *Bombus sonorus* is a furry mammoth of a bumble bee. A male *Hypomacrotera*, a subgenus of *Calliopsis*, has black wing tips, a sweet face, and the fuzziest little abdomen. *Leiopodus*, *Ptilothrix*, *Tetraloniella*, all exotic-sounding bees next to the more common *Bombus*, *Osmia*, *Xylocopa*, *Melissodes*.

I also eavesdrop when I'm in the lab.

I'm only at the Bee Biodiversity Initiative gathering for a few days, but while I'm there four Bee Course alumni, Joan, Denny, Don, and Tim, are there as well as two instructors and Jerry's assistant, Corey, also a Bee Course alumnus. Joan works with bees at the University of Massachusetts. Don works at the Smithsonian but as a lepidopteran guy not a bee guy, and Denny and Tim are amateurs who do volunteer work with bees.

As I work and listen, I overhear talk of bee trades, cursing and wrangling over locations and label making, fear of running out of pins, teasing. Don gets some grief for being a novice at bee bowling. Joan tries to pawn off some of the hundred plus little identical bees she caught in her bee bowls. No one is taking.

Where the bees are

Bee diversity is greatest in warm, dry, temperate climates. Yet for most organisms, the tropics are where the greatest diversity is found. Why not for bees?

Charles Michener, the father of bee taxonomy, speculates on this in his book *Bees of the World*. He notes that many bees nest in the ground and make cells that have fairly skimpy linings. Couple this with highly perishable baby food and a humid environment, and the situation is ripe for fungal attack. Also, eggs are usually laid right on the pollen-nectar mix. In a humid environment, the food might absorb water and the egg or larva might sink in and drown. How's that for a nightmare scenario? Bees that do tend to do well in the tropics are the ones that have fancy, well-coated cells or that don't nest in the ground.

Arizona has the right climate for high bee diversity as well as two rainy periods, mid-summer and winter, with different bees associated with each. In one thirty-five-mile-long valley on the Arizona–Mexico border near the Southwestern Research Station, 435 different bee species have been found. For comparison, the entire state of Florida has around 315 bee species, Maryland around 400. Of the 435 bees found in that valley, 223 have been found nowhere else (yet). Few places have been as well sampled as that valley.

It makes you wonder how many other bees might be out there if only we had the resources to look for them.

On my first night I catch a snippet of conversation.

"Oh my god," someone says, "Is that *Perdita minima*?"

"That's pretty friggin' small."

"I didn't even think it was a bee."

More background talk, then a pause.

"Terry, I have a question for you. I snapped its head off."

Whoever had beheaded the bee had turned to Terry Griswold for help. It was a logical thing to do. Terry is an instructor at the Bee Course and works at the Bee Lab in Logan, Utah, the rest of the year. Terry's work is largely systematics, who a bee is and where it fits in the bee family tree,

This *Centris rhodopus* male is from the desert near Palm Springs, California. *Centris* females often collect oils from plants that they either mix with the larval pollen wad or use in nest construction. ◄

Diadasia rinconis lives in a warm, dry climate as well. This male is a fluffy little bear of a bee from Tucson, Arizona. ▲

and he spends a lot of his time hunting for and identifying bees. When other people are trying to identify western bees and have whittled them down to the last recalcitrant few that just can't be identified, they head to the Bee Lab's large reference collection and Terry. The man has pinned and repaired many bees over the years and knows how to deal with a knocked off head—even a really tiny one.

I want to talk with Terry about being a taxonomist. Ever since I met eighty-something-year-old Robbin Thorp—who despite being retired has a giant backlog of bees to identify—I've wondered who will be identifying bees when Robbin is no longer around. It certainly seems like the academic world isn't overwhelmed with ID people or Robbin wouldn't be so busy.

Terry has a smile that pulls you in. It's both joyous and youthful and doesn't quite fit with his gray hair. While we talk, he pins, holding the bee between his fingers and popping the pin through its thorax. Bee after bee, zip, zip, zip, all pinned nice and level. After the many thousands of bees he's pinned, he clearly doesn't need to apply all of his concentration skills to getting it right.

One of the questions I've wanted to ask Terry is if he thinks taxonomists are a dying breed. When I ask, he laughs, "You're looking at it." He adds that the problem isn't that there's no one learning taxonomy anymore—he's had some great students—but that there are no jobs. At the moment, taxonomy isn't a subject that is valued, although it has been in the past. In the 1950s and 1960s huge numbers of bees were found and cataloged. Now people who need bees identified for pollination projects often don't have the ability to identify them, certainly not to the species level, yet they often don't budget for bee identification in their grant proposals. Terry admits that sometimes bees in those studies just get lumped together, "honey bees, bumble bees, and the little green ones." Clearly, that's not good enough for some projects, so people send their bees to Terry, who gets to them when he can.

When I ask him to estimate how many bees he identifies in a year, he pauses. "I don't know." Another pause, "Thirty thousand or so?" Somehow, supply and demand isn't working in the usual way here.

As Terry and I talk and he keeps pinning, zip zip zip, the talk and teasing of the others flow around us. I can't remember if it was the first or the second night that I saw the bee that totally captivated me. Denny had caught it, I remember that. It was a tiny bee and very rare, a member of the genus *Neolarra*. This little bee started me on a trip into a surprising aspect of life in the nest, one where battles to the death occur. It also gave me some insight into why Jerry Rozen has spent so much of his life sitting on the side of a hole with a blow tube and a penknife.

Neolarra and the cleptoparasites: death and thievery

The tiny bee flits about, just above the ground. Clothed in a tightly pressed suit of gray hairs, it is scarcely visible—a ghost bee. Of course, given that the bee is maybe half the size of an uncooked piece of rice, it could be bright red and still go unnoticed. The little bee may rest in the heat of the afternoon, head down on the shady side of a stem. At some point, if the bee is female, she stops flitting, having found what she's been looking for. She dives into the hole of another minuscule bee, one of the large genus *Perdita*. The goal of the ghost bee is theft and ultimately death. Her plan is to hide her egg in the cell of the host bee. Her offspring will hatch, kill the host bee's young, and consume all the provisions laid down by the hard-working host mother. North America may have a dozen or more species of *Neolarra*. All are cleptoparasites, those who thieve and kill rather than gather and provision.

I put together this description of *Neolarra pruinosa* based on a paper written in 1965, not long after I was born. I've never seen a *Neolarra* of any species flit about or hang from a stem. I've only seen one dead on a pin, but I still find it mesmerizing in its smoky grayness. I wonder what this little *Neolarra* and all the other cleptoparasites do in the dark of the nest. Fortunately for me, every day I'm at the Southwestern Research Station I find myself in the car with the guy that wrote that paper, Jerry Rozen, so I'll get my chance to find out.

Neolarra vigilans is a teeny bee, smaller than a grain of rice.

Nobody likes the idea of parasites. Thoughts of ticks and nasty intestinal worms tend to come to mind. But parasites are everywhere, and scientists have found that, generally speaking, they are beneficial. Not for the individual dealing with the parasite, naturally, but for the ecosystem as a whole. Usually, we think of parasites as evil outsiders. Bees are attacked by those kinds of parasites, like the vile *Varroa destructor* mites that destroy honey bee colonies, but about fifteen percent of bees are parasites of their own kind. Somehow, that seems much ruder than being preyed upon by some Other. You don't expect your cousins to turn on you.

The nature of bee parasitism is surprisingly variable. A particular parasitic bee is often picky, going after one bee species or a small group of closely related bees. For bumble bees, the primary method is for the parasite to kill

Members of the genus *Triepeolus* are cleptoparasites of various kinds of bees.

off the queen, take her place, and have the former queen's daughters raise her eggs. I often see them called cuckoo bees, whereas the bee parasites of solitary bees are usually referred to as cleptoparasites—thieves *and* parasites. My adorable little *Neolarra* is one of these thieving killers.

One day, as we are driving, I ask Jerry about *Neolarra*. "How does it attack the *Perdita* [its host]?"

"It doesn't attack the *Perdita*," he responds. "It enters the nest of the *Perdita* when the *Perdita* female isn't there."

Jerry admits that he hasn't worked much with *Neolarra* (and his paper on them *was* written more than fifty years ago), but he assumes it acts like most parasitic bees. He goes on and tells me about *Oreopasites*, a close relative of *Neolarra*. (I put together the following description from what Jerry tells me in the car and some research papers.)

A cleptoparasite female, with eggs ready to lay, scouts for a likely nest entrance. At least some seem to recognize where they need to be by smell. The cleptoparasite mom flits around or finds a handy perch and keeps an eye on things. She sees a bee come out of a hole. Here's her moment. Provisioning a nest cell for an egg usually takes multiple trips, so chances are the cell is open and partially full. The clepto mom zips in, finds the partially filled nest cell, and usually hides the egg in some way.

According to Jerry, the egg is usually partly buried in the cell wall. If the host bee comes back and finds it, she'll dig it out with her mandibles and kill it. I confess that I'm surprised the host bees don't routinely find and kill these eggs, given the small size of the nest cells. Jerry replies, "They have to find the egg. And you know what? They don't have a flashlight."

Other kinds of parasitic bees do things differently. They don't look for a partially provisioned cell, but instead hunt for one that's just been closed: no host mama bee to return and potentially find the hidden egg. These cleptoparasite moms make a hole with their mandibles big enough to stick their abdomen in. Then they lay an egg and cover their tracks. A few parasitic mama bees kill the host egg, but most often it is up to the young larva to deal with the host bee baby themselves, and these parasitic larvae are built for killing.

When most bee larvae hatch, they are helpless white grub-like things. They lie on their food, eat, and grow. Because they are insects, they periodically molt (shed their exoskeleton). Each stage between a molt is called an instar, and bees generally have four or five of them. The first instar of most parasitic bees emerges looking very different from a non-parasitic bee. It is not a helpless white grub-like thing. It is a white grub-like thing with sharp mandibles and strong head muscles, the better to bite and kill with. Cleptoparasites are said to be hospicidal—host killers. In addition to the physical equipment needed to slay the hosts, they also need a hospicidal nature.

According to Jerry, some parasitic bees attack the host as soon as they get their head out of the egg, sucking the life out of the host egg. Then they can finish hatching at their leisure and eat up the pollen reserves meant for their fallen foe. Most cleptoparasites, however, wait until they've completely hatched before dispatching their cell mates. The extra strong jaws and mandibles aren't really needed after the parasite has killed the host, so the following instars usually look more like a normal bee larva. Sometimes more than one cleptoparasite mom lays an egg in

Stelis laticincta, here in Mendocino County, California, is one of the thieving parasites.

a cell, so not only does a parasite larva kill the relatively helpless host, it may have to do battle with others of its kind. Presumably, the first one to hatch has a big advantage here.

Sometimes though, it's not the first instar that does the killing.

The cleptoparasite *Stelis ater* may kill the host at any stage in its development. It all depends on when it encounters the host bee, because neither larva can move. (Many cleptoparasite larvae can.) In one study of these bees, the researchers found a fifth instar cleptoparasite munching away on the pollen wad right next to the fourth instar of the host. About an hour later, the cleptoparasite had clearly found the host because it had taken to chewing on it instead. Hospicidal nature indeed.

Even more surprising is little *Leiopodus singularis*, a cleptoparasite of *Diadasia olivacea*. Researchers observed these bees happily going about their business right at the Southwestern Research Station. They found that *Leiopodus singularis* would often perch outside the nest area of *Diadasia* with an air of "furtive alertness." Sometimes a *Leiopodus* female would go down a hole and be back in short order, maybe twenty seconds. Other times, the bee was gone for around two minutes, and the researchers

Holcopasites calliopsidis is a cleptoparasite of bees in the genus *Calliopsis*. This one was found in Pickens, South Carolina.

assumed that during these longer visits she laid an egg. After these visits, most of the *Leiopodus* bees would perch and preen for about thirty seconds, then fly off. I thought this was a "Yay, go me, I laid an egg" kind of preening, but the researchers thought it was likely to be merely cleaning up after being in a strange nest. I suppose it's like washing your hands after touching someone else's stuff.

Where things get really interesting with this bee is with the larvae down in the nest. *Leiopodus singularis* has a long incubation period. When it hatches, the puny little one-millimeter first instar is confronted with a gigantic third or fourth instar (ten to twelve millimeters) of the host, which it promptly kills. That little instar has fangs and is a mighty crawler, despite having no legs. Somehow those two features, and presumably a generally murderous nature, allow it to dispatch its massive foe and eat up the rest of the food provisions.

When Jerry told me this story, I asked how on earth he knew that the first instar killed the fourth? He'd dug up nests, and there they were: little first instars sharing a cell with their slain fourth instar rivals.

This *Nomada affabilis*, an exotic-looking cleptoparasite from the wilds of Maryland.

Bees often seem to like sleeping upside down. This female of the cleptoparasitic species *Coelioxys rufitarsis* is sleeping in the bee garden at the University of California Berkeley.

The state of bees in the wild

I'd thought this chapter was going to be about the importance of wild bees, in a big, arm-waving kind of way. Then I got shanghaied by that ghostly little *Neolarra* and its murderous and thieving kin. Nevertheless, I can't completely ignore the bigger picture since around three-quarters of the world's flowering plants have animal pollinators, mostly bees.

We really don't have a good idea of how well wild bees are holding up to habitat loss, degradation, and fragmentation plus whatever effects climate change, imported bees, parasites, and diseases may be having. We do know that some bees are in decline. The data on the current status of most bees are patchy at best, and few areas have been well studied. Bee surveys take a huge amount of time and money, and someone has to identify all those bees. To make matters worse, in most places we have little or no historical data to compare current results to. Fortunately, in a little town in the hinterlands of Illinois we do have a small window into the past.

Between 1887 and 1916 Charles Robertson, a professor of biology and, oddly, Greek, studied the plant-pollinator interactions around Carlinville, Illinois, about 250 miles southwest of Chicago. Robertson was unusual for the time in that he didn't just collect the insects; he also noted which plants they were visiting and when the plants bloomed. Over those twenty-nine years, he counted 1429 pollinators (of all types, not just bees) on 456 plant species. A second smaller study done in the area in the early 1970s found few changes in the bee fauna since Robertson's time.

By 2009 and 2010 when a third study was done, things had changed. The researchers for this project looked at a subset of the plants that Robertson studied: the woodland spring ephemerals. A century earlier Robertson had found 532 plant-pollinator pairings among these woodland plants. By 2010, 407 of those pairings were gone. Many of the pairings broke because half the bees involved had disappeared from the area. Surprisingly, however, 120 new pairings had developed, demonstrating some resilience in the system.

Agapostemon sericeus

Between 1974 and 1980, *Agapostemon sericeus* nests were found and studied on the campus of Cornell University in western New York. These bees are solitary and seem to be easy-going in their nesting desires, although they clearly have a liking for lawns. Some dug their burrows in a little secluded patch of moist lawn surrounded by gardens. Others chose a lawn often walked upon by students. Some decided to set up in company with *Agapostemon virescens* in a weedy area that was seldom mown. Still others chose a well-drained lawn that was also inhabited by a large aggregation of other halictids and *Andrena*.

Agapostemon sericeus is a bee that was present in Carlinville, Illinois, a century ago that has disappeared from the area.

Part of the reason for the losses seems to be climate change. Winter and spring temperatures have increased by about 3.5°F (around 2°C). Both flowers and bees rouse earlier and have shorter seasons, but the changes in their seasons aren't quite in sync. Habitat change has also played a role. Much of the land that was forest and prairie in Robertson's time is now agricultural. Despite the changes, the plant-pollinator community hasn't completely broken. It has adjusted, but it is also more tenuous. The ability to keep readjusting isn't infinite. At some point, the resilience will give way. Tiffany Knight, a researcher from the 2010 study, said in an interview that we can't just keep kicking plants and bees and expect them to continue finding ways to function.

The American landscape is likely full of Carlinvilles, or places where things are even worse. We just don't know. The hopeful piece is in the resilience that the plants and pollinators have shown around Carlinville. If we start replacing habitat where we can, providing nest sites and floral resources, maybe we can stop the slide before we've kicked a good chunk of the bees to death. There's an easy place to start. Most of us don't have the ability to turn agricultural land back to woodland or prairie but ordinary people do have the power to transform one of the largest monocultures in the country into a place for pollinators. We can bring flowers back to the lawn.

Bees in the Grass: Rethinking Normal

NESTLED IN THE HOLLOW below me is a small patch of plants. I'm going to call it a meadow because I've traveled 3000 miles to see it and calling it what it looks like, a good-sized patch of weeds, would make my trip seem foolish. On this late August day, few flowers are on display. From a distance, the patch appears empty of life. Yet as I walk closer and then wade in among the tangle of stems, my perception changes. This plot of ground teems with life. A large clump of goldenrod brightens one end, not too far from a bluebird house. Stems of scarlet blanketflower mingle with those of starry lavender asters. A few small, yellow daisy-like flowers (the type that are generally reviled by people trying to identify them and referred to as DYCs for "damn yellow composites") froth around at ankle height with a lonely red lobelia. Dominating the patch is spotted bee balm (*Monarda punctata*), a wildflower native to eastern North America. The bee balm is hip high and stands up straight, with no lazy leaning. Its color scheme is subdued yet pleasing, and dozens of big black carpenter bees buzz among the flowers. The bees perch on the washed-out pink bracts (leaves modified to look like petals) with their heads stuck deep in the burgundy-spotted yellow flowers, feasting.

The day is overcast and moody, but I stand among the plants for a good fifteen or twenty minutes. I see a really cool wasp and at least six kinds of lepidopterans. (I use the collective term for butterflies and moths because I often can't tell the difference between them. But for me, coming from butterfly-starved Seattle, they alone make the trip worthwhile.) The young from the bluebird house have fledged. The mason bees who flew earlier in the season have filled the holes in the nest block, the mason bee babes within tucked in for their winter's sleep. I've been told that sweat bees had flown earlier in the season. I assume that their babes are also snoozing away, waiting for spring. I don't see any of the three pairs of hawks that live in the area. I watch the bees in the quiet little patch and every now and then hear a *thwack* or the hum of an electric motor, the sounds of golf.

In 2015 the United States had 15,372 golf courses, averaging around 150 acres per course, which comes out to 2,305,800 acres. We've got the equivalent of a Delaware (1,296,640 acres) and a Rhode Island (775,680 acres) devoted to golf courses in this country, and a lot of that land isn't even used to play the game, unless the golfer messes up. Golf courses have an in-play area (the tees, fairways, and greens) and an out-of-play area (the rough, bunkers, water hazards, parking lots, buildings, and the like). The in-play area may be as little as forty percent of the course or as much as seventy. Currently, a lot of that out-of-play area is turfgrass, but it doesn't have to be.

Several groups and programs are working to make golf courses more environmentally sound. In 2002 the Xerces Society for Invertebrate Conservation, in conjunction with the U.S. Golf Association, wrote a report called "Making Room for Native Pollinators: How to Create Habitat for Pollinator Insects on Golf Courses." Audubon International (no relation to the Audubon Society) has developed a sanctuary program and golf courses can apply for certification. More recently, a new program has come to the United States from Europe, Operation Pollinator, a program started by Syngenta, one of the world's largest agrochemical companies. Golf courses and a pesticide company collaborating to build habitat for pollinators? What's the catch? I wanted to know more, and a lot of the Operation Pollinator–golf course connections I found were in the Deep South, so I decided it was time for a trip back home.

Operation Pollinator

Walt Osborne works for Syngenta, and one of his duties is talking to people like me who are interested in Operation Pollinator. I wanted to learn how and why a company that makes products that deal wholesale death to some insects decided to start a program to save others.

Walt told me that Operation Pollinator started in the United Kingdom in the early 2000s as the Buzz Project for farmland and then spread to other countries. Along the way, the name changed and it got taken up by golf courses. He told me that the move to golf courses came about when two different pieces of information made their way to the folks at Syngenta. Surveys of golfers had found that one of the things they liked about golf was the nature. Also, golf courses were looking for ways to cut costs. The people at Syngenta mulled over what they could do to help their golf clients. A lot of the out-of-play areas on golf courses are managed as turf, but what if they weren't? The turf people at Syngenta talked with the flower people and the bug people and came up with the idea of turning some of that grassy out-of-play area into meadow—making pollinator habitat. This plan would cut down on costs and provide more nature for the golfers, and so began Operation Pollinator for golf courses.

Distrust of agrochemical companies runs high for many, myself included, and I still wondered where the catch was. So I asked. Walt said that Syngenta is not just about making money, and that's one of the reasons he likes working for them. Sure, Operation Pollinator has a public relations component to it, but he said the program was never about the bottom line. It was a chance for all the horticulture types and biologists that work for Syngenta to use their expertise "to do the right thing." I still keep suspecting some Machiavellian plot, but maybe I should let it go. Regardless of why, Operation Pollinator has led, and continues to lead, to pollinator habitat being developed—on golf courses of all places.

Operation Pollinator for golf courses came to the United States in 2012, and by 2016 more than 200 golf courses in twenty-nine states had an Operation Pollinator plot. The plots range from half an acre to more than a hundred. Syngenta provides how-to information, signage, and some help with ways for golf courses to get the word out. The first Operation Pollinator plots in the United States were started as part of a scientific study by Emily Dobbs, a graduate student at the University of Kentucky, and her advisor, Dan Potter, at five golf courses around Lexington. Their goal was to conduct a two-year study to determine the best way to manage these kinds of plots, assess floral resources throughout the season, and see which bees came to visit which plants.

Emily and Dan used three seed mixes (about twenty-five species total) that were specifically developed for the Operation Pollinator plots in Lexington. They prepped the ground, which in the model given by Operation Pollinator has you killing off grass and weeds with Syngenta products, and planted.

The plots changed between the two years, as did the bees that visited. Some of the plants that thrived the first year languished the second, while others came on strong. Only two species of bumble bees showed up the first year but seven, including three declining species, came to dine the second year. In both years, halictids (sweat bees) were the most ubiquitous bees, both in number of species and number of individual bees. *Halictus ligatus* won the award for most bees caught in both years. In 2012, the researchers netted 232 individuals of this species—on only three plant species—ten times more than any other bee species. The second year slightly fewer *Halictus ligatus* were caught and some bumble bees made a strong showing, so the *Halictus* only managed to be three times more abundant than the next most plentiful bee.

Ah, those halictids. This isn't the first time I've seen one of them completely overrun all the other bees. What makes them so ubiquitous?

Fat mamas: *Halictus ligatus* and the sweat bees

Halictus ligatus belongs to the Halictidae family, which are called sweat bees because of their sweat-licking tendencies. The family seems to be exceptional in its variability and adaptability. Halictids' appearance ranges from the glorious iridescent green of *Agapostemon* to the dowdy. The genera *Lasioglossum* and *Halictus* contain a plethora of dark bees whose subdued appearance seems made for skulking about unnoticed.

Excess seems to be part of the nature of halictids. The family has around 3500 species that span the globe. Members of the genus *Lasioglossum* win the prize for most species, with around 1800. Some of the halictids are also remarkably adaptable, encompassing huge and variable ranges. *Halictus ligatus* can be found from northern Canada all the way down to Venezuela. Another *Halictus* species is able to leap oceans and mountain ranges, being found in Europe, northern Asia, and across the breadth of the United States and Canada. These are not finicky, princess-and-the-pea kinds of bees.

A particular location likely won't be overrun with dozens of species of halictids, but a few of the species that are there are likely to be very successful in terms of numbers. Why a bee is abundant may have a lot to do with lifestyle, and here once again the halictids show diversity. The family Halictidae covers almost the entire behavioral spectrum from solitary to social. Some individual species are even flexible enough to vary in their sociality depending on where they live. In the high mountains where the flying season is short, bees of a particular species may be solitary, whereas their relatives living at balmy lower elevations may be primitively eusocial (overlapping generations and cooperative nesting). Some of the social halictids produce several generations a year and so the number of bees just keeps building over the season. In the processing work I did in Seattle, the number of halictids found in the bee bowls grew from "We've got some halictids" early in the season to "Please no, not more halictids" by midsummer.

Halictus ligatus, the bee that dominated the Operation Pollinator trials in Kentucky, is primitively eusocial. Female family members live and work together in the same nest throughout the season, with one queen to rule them all and lay the eggs. As with honey bees, what a female *Halictus ligatus* bee larva eats influences what kind of bee it grows up to be. In the honey bee world, a female's place is set absolutely by the food she gets as a larva, but some flexibility is built into the *Halictus ligatus* world.

Not all *Halictus ligatus* look like they've been rolling in a pollen bath, but this one certainly does. ▲

A little *Lasioglossum* sweat bee clings to *Monarda*. ◄

Lasioglossum paraforbesii, a somber halictid with only a hint of amber in the wings to liven it up, is a fairly uncommon species of a ubiquitous genus. This bee is from the Badlands in South Dakota. ▲

An *Agapostemon texanus* female leaves her nest in El Cerrito, California. Who wouldn't want to find this climbing out of a hole in their yard? ◄

Some female eggs get bigger pollen-nectar wads than others, leading to bigger bees with more fat stores. These fatter bees can immediately go into diapause, digging themselves sleep cells in extensions off the nest in which they were born. They sleep through the winter and are able to start a nest and rear young the next year. Their smaller, leaner sisters have to go out and forage to build up sufficient fat stores to overwinter and can't immediately go into diapause. Because the existing queen is bigger than these lean bees, she can harass them and try to prevent them from digging a sleeping place once they've fattened up. Even though some of these smaller females have mated and are thus able to start a viable nest, the queen may be able to persuade them to stay in her nest and act as foragers instead. They give up their hope of offspring and become worker bees. When it comes to *Halictus ligatus*, the big fat bees usually get to be the mama bees.

Back on the golf course

Golf is loved by some and reviled by others as an elitist sport for white guys. I have never played a round of golf, although I had quite a bit of exposure to it as a kid. From the fourth grade on, I lived in one of those suburban neighborhoods built around a golf course. My brother has played golf all his life. My dad, who never played, would regularly watch golf on TV. On weekends as I sat in the corner of the couch reading my Nancy Drews, the quiet sound of golf commentary flowed around me and the names wafted into my consciousness: Arnold Palmer, Jack Nicklaus, Lee Trevino. I took lessons briefly as a teenager, to keep my aunt company. It didn't take.

I've never lamented my lack of golf experience, but plenty of others feel the pull of the links. The National Golf Association reports that in 2014 24.7 million people in the United States (over the age of six) played one or more rounds of golf during the year. At least 2 million golfers are serious enough about the sport to have a handicap. These folks average thirty-seven games (men) and twenty-two games (women) per year. So, you figure that during the good weather, a serious golfer goes out about

The power of scent

Scent is used by bees in a wide variety of ways:

- Some *Nomada* males practice olfactory trickery. To aid their cleptoparasitic mates in their thieving ways, the males spray the females during sex with a scent that smells like the bee they are about to parasitize.
- Some bumble bees and honey bees mark flowers with a scent to show they've recently been pillaged and aren't worth another visit right away.
- A variety of bees secrete pheromones that are attractive to potential mates.
- Other scents are meant to be turnoffs. *Lasioglossum zephyrum* males are attracted to the scent of virgin females. Once females are mated, though, males lose interest in them. Research has shown that males leave behind a chemical that wards off other suitors—a bee anti-aphrodisiac.

Nomada species are cleptoparasites that usually go after *Andrena* mining bees, but not *Nomada articulata*. This little beauty goes after other beautiful bees, *Agapostemon*.

This *Nomada* bee looks innocent as it visits the flowers of *Melilotus officinalis* (yellow sweet clover) in Stanislaus County, California.

once a week and spends around four hours whacking, hunting, and chasing a little ball. Why?

The reasons seem to be many: the chance to play with old friends or to meet new people or to have a few hours of solitude. The satisfaction of a well-hit shot shouldn't be underestimated either. Golf is a sport that can be played for a lifetime and anyone can play, even people who aren't great athletes. Tucked away amid all the other reasons people like golf is the fact that it happens outside, in nature. Given all of the inputs that usually go into a golf course—water, fertilizers, fungicides, insecticides, herbicides, gas for equipment—some people may view a golf course as anti-nature. Yet a golf course is still a large area of green growing things. Of course, some golf courses feel more natural than others.

Mark Hoban is the golf course superintendent at Rivermont Golf Club in Johns Creek, Georgia, near Atlanta. He has developed a golf course that feels—not wild, exactly—but like a really nice park. Trees are plentiful, and he's added native plants that color up nicely in fall. I haven't been to many golf courses, but I know they aren't all like this. I visited one that felt like a symphony in turfgrass—or maybe a string quartet—with grass in short, medium, and long with a few trees thrown in. Not only does Rivermont feel like nature (of a well-groomed sort), it actually is a lot closer to nature than most golf courses.

The course is run mostly organically. Mark has not just a worm bin for making compost, he has a multilevel worm condo and feeds the worms on the leavings from the clubhouse's kitchen. He brews compost tea. Since Mark started using his homemade compost products, he's reduced Rivermont's chemical nitrogen use from two and a half pounds per thousand square feet to less than half a pound. He uses insecticides only as a last resort and as spot treatments rather than proactively on a schedule like many golf courses. He relies instead on biologicals and beneficial nematode inoculations. The golfers seem happy with the course they have, and part of that is the nature. Mark told me he's had golfers come in on days when they weren't playing golf because they'd taken on a bluebird house as their responsibility and they wanted to check on their birds.

Rivermont Golf Course in John's Creek, Georgia, has the feel of a park.

Research has revealed that golf courses can be better than surrounding green spaces for an array of animals: various birds, reptiles, frogs, and, of course, insects. Rivermont really does look like a place where a lot of wildlife might live. Plentiful trees, a diversity of wild grasses, and that little Operation Pollinator patch play host to a surprising number of animals, and not just for a month or so. The Operation Pollinator patch alone supplies food for pollinators throughout the entire growing season, which is a prime objective of good habitat. Next to the planting is a muddy ditch that provides building material for mason bees, puddling spots for butterflies, and water for all.

On the day I visit, it happens to be big old carpenter bees (*Xylocopa*) who are taking advantage of what's on offer, plundering the native bee balm. I'm sorry to have missed some of the other bees, but it's good to see someplace where large carpenter bees can live safely, because they are bees that are frowned upon by many.

Aerial patrollers: *Xylocopa virginica*, the eastern carpenter bee

Xylocopa females use their extraordinary mandibles to chew into hard fresh wood to make their nests. In *The Bees in Your Backyard*, Joseph S. Wilson and Olivia Messinger Carril refer to mandibles as "the bee equivalent of the opposable thumb." To me, mandibles seem at times like a cyborg combination of hand and tool. Where *Xylocopa* bees are concerned, their magical mandibles are the cause of the dislike felt by some toward these bees, because sometimes the hard fresh wood that carpenter bees burrow into is part of a house. *Xylocopa* bees usually avoid painted wood, but any unpainted wood, whether fence post, porch rail, or eave, seems to be a fine place for them to make a home.

Bees in the genus *Xylocopa* are big and so make big burrows. They generally dig straight in, leaving a half-inch-wide hole, and then turn, excavating along the grain of the wood. The bees sometimes reuse old burrows, but they often seem dissatisfied with them and feel the need to do some cleaning and remodeling by excavating and extending the tunnel. Completion of the tunnel doesn't mean the end of the digging either. After a *Xylocopa* female provisions a nest cell, she gouges bits out of the walls of the burrow and glues the wood scrapings together to make a little particleboard barrier between the cells.

Only the females burrow into wood, but *Xylocopa* males also do their bit to annoy homeowners, buzzing people and invading their personal space. Perhaps *Xylocopa* males can be forgiven their behavior when you realize what's driving them.

A lot of bee scientists focus on female bees. They lay the eggs, build the nests, gather the food stores, and do the bulk of the pollination. Often, the males seem to be little more than sperm donors. Nevertheless, some of those males have pretty interesting behaviors too. Of course, the interesting parts are usually related to sex in some way: hoping for sex or waiting for sex or having sex. Some bumble bee males congregate like guys at a bar and fly about together, waiting for a willing female to come along. For other genera, males are not nearly so friendly and social and have a territory that they guard with great vigilance. *Xylocopa virginica*, like those found at Rivermont, fall into the territorial group, but they go about it a bit oddly.

Xylocopa virginica males that dive bomb or buzz people are just doing their job, guarding. These males can't sting, but I'm sure it's still unnerving. The guys may decide to guard a variety of places, often an area near a nest or a food source, but sometimes they'll choose something to guard that seems a little strange, like a rock.

Edward Barrows watched male *Xylocopa virginica* over the course of several years and noted that a number of males had set up territories near a nest. They had a hover zone that was usually only about a cubic foot, although variously shaped, perhaps even bending around a corner. The males would attack anything that seemed bee-sized out to a zone of sixty feet or so. A few *Xylocopa* actually flew at an airplane that was far enough away to look about the size of a bee. They didn't go after low-flying and therefore "bigger" planes, so they weren't just driven to great acts of stupidity by some bee equivalent of testosterone poisoning.

Barrows found that some *Xylocopa virginica* males would set up territories right next to each other. They seemed to have some gentleman's agreement going on, because all the bees carefully ignored the guys in adjacent territories while attacking others who came in from elsewhere. When territorial disputes broke out, the males would chase each other, pouncing and sometimes knocking each other out of the air. When a bee would leave his territory for a bit—a guy's gotta eat—interlopers would sneak in and try to mate with the females in the area only to be chased away by the returning master of that territory.

I've always thought of male bees as lazing about a lot, waiting for a chance to have sex while the females did all the work. Clearly, trying to have sex *is* a lot of work for *Xylocopa virginica*.

Lawns in America

Pollinators—butterflies, wasps, small innocuous bees, big potentially annoying bees—all living in habitats built by golf courses, supported by a pesticide company, and approved by golfers. All it took was a shift in

Xylocopa virginica males have great big eyes, the better for seeing all those potential invaders.

mindset to create a new normal. What golf courses are doing with Operation Pollinator is going a step beyond just adding some flowers to the grass. They are removing turf in areas that don't need to be grass and replacing it with flowers for pollinators. That's one approach, and it works in some places. Sometimes, though, you want a lawn to be a lawn, a place for play, picnics, and soccer pitches. What if a lawn can be all that and a place for pollinators too? The continental United States has 30 to 40 million acres of lawn—acres filled with possibilities. If golfers and golf course superintendents can rethink normal for golf courses, then surely the rest of us can reimagine our lawns as a place for grass *and* flowers. After all, that's the way it used to be.

Some gardeners are violently opposed to lawns. I'm not. Used strategically, a flat swathe of green can do a nice job in setting off garden beds. Plus, it gives the kids a place to play. Nevertheless, if you think about it, a lawn is an odd thing. It's like we are trying to create a rug, something static and unchanging, out of a group of plants. Also, with all the plants we can choose from, we decide to devote a significant part of our yards to a monoculture that produces absolutely nothing. Our lawns soak up a lot though: money, fertilizer, water, herbicides, insecticides, time. A lawn does have one significant virtue: it can be walked on. A *good* lawn can even be walked on in the rain without getting your feet muddy. Some of the other attributes a lawn may have, like slowing surface water runoff or protecting soil from erosion, can likely be done better by other plants. In some places a lawn's walkable, no-mud attributes make it the right choice for the spot, but most lawns are just placeholders. How did this come to happen?

At least a part of the allure of lawns seems to be of the keeping-up-with-the-Joneses variety and always has been. Lawns have been around since the Middle Ages. In the twelfth century, King Henry II of England was said to have lawns at one of his palaces. The lawns of his grandson, King Henry III, at the Palace of Westminster were apparently rolled and mown—lots of work—which sounds like a lawn alright. By the eighteenth century in England, not just kings but regular, rich lordly types had lawns around their country estates, which were shorn by men with scythes and/or grazing animals. Keeping a lawn short and looking good in those days was highly labor intensive, unless you went the sheep route. At Blenheim Palace in England it took fifty scythesmen to cut the lawn. Every ten days or so they'd line up and march in rows, scythes swinging. Later in the day the women and kids would come out to collect the cuttings. Wealthy Americans who visited Europe saw some of these lawns and worked to recreate them in the New World.

Over time, the idea that a lawn is a desirable feature filtered out to a wider and wider audience. Having a good lawn got much easier after World War II thanks to the widespread availability of herbicides, insecticides, synthetic fertilizers, better mowers, and improved grass cultivars. At the same time the land of lawns, the suburbs, was growing. Right after the war the first mass-produced suburban community was built: Levittown, New York, 17,000 homes, all with a lawn.

For some, there seems to be a competitive edge to lawns, the search for perfection, with perfection equaling a uniform green sward free of life other than the one species of grass that has been chosen to grow there. Lawns weren't always like that. Once upon a time they had flowers—and life. What if we turned back the clock, let some flowers back in? We could start easy. We don't even have to work at letting a really good bee flower into the lawn. We just need to ease up on killing it. Let the clover return.

In defense of clover

Clover is a wonderful helper for lawn grass. Where clover and grass intermingle, the grass grows lusher and greener. Grass loves nitrogen, and clover houses bacteria on its roots that snatch nitrogen out of the air, turning it into a form that can be used by plants. It's actually pretty magical. Some of that nitrogen gets used by the clover itself, but some ends up in the surrounding soil, available for uptake by those nitrogen-hungry grass roots. Because we routinely chop off part of the grass plants that make up our lawns and haul those chopped bits away, any nitrogen that was in the scraps can't be recycled. Letting the clippings lie where they fall allows some of that nitrogen to be returned to the soil, but nitrogen is a fickle and ephemeral nutrient, constantly mutating forms and disappearing from the root zone. More is always needed.

Grass seed mixes used to contain clover seed to help meet some of those nitrogen needs. Studies have shown that if a nitrogen-fixing legume, like clover, makes up about thirty percent of a grassy area, it supplies most of the nitrogen the grass needs. That's nice for the grass, and the reduced fertilizer need saves money and is good for the environment. Clover does more than suck nitrogen out of the air and convert it into a plant-friendly form; it also flowers. Those flowers are a food source for a whole variety of pollinators.

A study done in Lexington, Kentucky, looked at the pollinators on two common lawn weeds: dandelion and white clover. On clover they found twelve bee species as well as ten species of other pollinators. The dandelions had twenty bee species and five other pollinators. Ian Lane, a graduate student at the University of Minnesota, counted bees visiting clover in lawns and parks around Minneapolis. He found thirty-seven different species feasting away, including one with very specific pollen needs, *Calliposis andreniformis*.

Bees in the genus *Calliopsis* belong to the mining bee family, Andrenidae, so it's no surprise that they dig holes in the ground for their nests. Many of these bees are pollen specialists. Although pollen specialists will stop and gather nectar to refuel from plants other than those they specialize on, they will collect pollen for the babes only from their special plants. *Calliopsis* species get placed into subgroupings depending on what kinds of plants they like, and *Calliopsis andreniformis* is in the group of legume lovers.

Calliopsis species seem to be fairly uncommon, except when they're not. Of the thirty-seven bee species found on Minneapolis clover, *Calliopsis andreniformis* was the fourth most frequently seen. These bees are called campus bees, perhaps because they were found on some campus with clover-infested lawns. Clover is certainly something *Calliopsis andreniformis* likes—a lot. A study that looked at the pollen types collected by this bee found it was almost completely clover pollen, albeit two different kinds.

The males of this species seem like they'd be fun to watch. They're patrollers. When an interloping male enters the territory of another *Calliopsis andreniformis*, the males face off. They start flying upward in a tightening spiral. Sometimes the invader flies off, but other times both bees fall to the ground, where they wrestle and bite each other's legs. Mixed martial arts matches in the bee world.

When I first started learning about bees, someone told me that one wouldn't expect to find specialist bees in urban areas. Yet here is one happily gathering from a lawn weed in Minneapolis alongside thirty-six other species of bees. Perhaps it's time to change the designation of clover from lawn weed to pollinator plant. Some scientists in Minnesota are hoping to persuade people to do just that.

The making of a bee lawn

Mary Meyer, a grass person, and Marla Spivak, a bee person, both at the University of Minnesota, brought their expertise together to develop a

Bombus griseocollis the brown-belted bumble bee, has a range covering most of the United States and part of Canada. It and several other bumble bees were found visiting clover in lawns in both Lexington, Kentucky, and Minneapolis, Minnesota.

lawn that would work for bees *and* people. They call it a bee lawn. The goal is to have a lawn that functions and looks, mostly, like a normal lawn but also provides flowers for bees.

Meyer and Spivak aren't the first people in recent time to think of putting flowers in the grass. Back in the 1990s several companies came out with grass mixes that had flowers. Their goal wasn't to help pollinators but to make lawns less needy, requiring less water, fertilizer, pesticides, and mowing. Those mixes had names like ecolawn or fleur-de-lawn. The flowering plants weren't chosen with bees in mind, and, depending on the mowing regime, they might not even get a chance to form flowers. Meyer and Spivak wanted a lawn where the flowering plants were actually able to bloom, which isn't that easy since they are constantly being beheaded. To come up with a solution, they did what professors usually do: they applied for a grant and put a graduate student to work on the problem. Ian Lane, who did the clover study, got the project started.

Andrena wilkella was the third most common bee found in the study of bees on clover in lawns around Minneapolis. ▲

Check out the eyes of this *Calliopsis andreniformis*. The compound eyes of bees can be surprisingly variable in color and appearance. ◀

Tickle bees in the grass

Adding flowers to lawns for bees to dine on is great, but allowing a few hundred to several thousand bees to live in the lawn is taking bee love to the next level. It happens. Dozens of *Andrena vicina*, a mining bee, nested in a suburban lawn near Seattle, Washington, for years. Thousands of so-called tickle bees (*Andrena* species) live in the lawn at Sabin Elementary School in Portland, Oregon. A video taken at the school shows bees zooming around the lawn as children play in the background. An estimated 20,000 *Andrena* bees nest in the lawn, and no one worries about them because these particular bees don't sting, not even the females. A sign set up in the grass proclaims "Sabin School, Home of the Tickle Bees." I'm always hoping to make bees seem less scary, and giving them a nickname like tickle bee seems like an excellent start.

An *Andrena subtilis* male takes a break on a leaf in Albany, California.

The crew at the University of Minnesota faced a couple of problems in developing their bee lawn. First, they needed the right type of grass, one that wasn't too aggressive and wouldn't outcompete the flowers. The grass also needed to be fairly slow-growing, so it wouldn't need cutting as often, giving the flowering plants a chance to actually make some flowers. They chose hard fescue (*Festuca trachyphylla*) because it met their requirements and works well across the northern plains states. Next they needed to find the right flowers.

They were hoping to use natives, but North American native plants don't like living in lawns. Unlike some of the Eurasian native meadow plants, North American natives haven't evolved with a bevy of grazers, like sheep and goats, regularly chomping them down. If something crops (by teeth or blade) most of our native plants down to a few inches, the plants usually just die. Ian Lane found that out when he started his plant trials. He picked sixty-four kinds of plants, both native and non-native. He planted them and nurtured them, and then he mowed them down. Great death resulted.

Those plants that made it through this first cut (ha!) also had to fulfill some other requirements. In addition to not flat out dying when they were mowed, the chosen plants had to flower regularly after they were mowed. They also had to be able to withstand a fair amount of close competition. The flower seeds needed to germinate in the lawn situation, and their roots needed to be able to compete with those of the grasses. Of the original sixty-four species, only three made it into the current bee lawn seed mix: *Prunella vulgaris* subsp. *lanceolata* (self heal), *Thymus serpyllum* (creeping thyme), and *Trifolium repens* (Dutch white clover). Only *Prunella* is a native. Ian does hold out hopes for a couple more species.

Each of the three flowers in the bee lawn mix has its own suite of bees. Honey bees don't visit the self heal at all, but they like the thyme and the clover. An array of long-tongue bumble bees think the self heal is splendid, as does a whole range of halictids. A lot of short-tongue bees and males went for the thyme. The clover had a variety of bees, including the legume specialist *Calliopsis andreniformis*.

The folks at the University of Minnesota recommend cutting the grass of the bee lawn no shorter than three and a half inches, which some may view as a bit long. Ian admits that when the grass gets to four and a quarter inches, cutting time, it starts feeling a trifle "unruly." The problem with going shorter than three and a half inches is that only the clover is likely to keep flowering.

Mary Meyer told me that in some respects the hardest part of getting a bee lawn into use isn't developing the seed mix; it's dealing with people's vision of what a lawn should be. She told me that she studied the ecolawn-style grass-flower mixes back in the 1990s when they first came out, and she asked people if they would have this in their yard. About half said yes and half said no. Mary was left with the impression that "If we didn't have to worry about our neighbors, I think there would be a much more diverse look."

It makes sense. If you have a seamless ribbon of grass front lawns and then one person throws it off with longer grass spangled with flowers, well, the neighbors might object. But what if everyone decided that flowers in the lawn were good and that monocultures that do nothing but hold a place are passé? Again, perhaps it's just a matter of retraining our eye. Rather than looking at a lawn and thinking "it should look like a golf course," we can aim for a vision closer to Julie Andrews and the "fields are alive" from *The Sound of Music*, only ankle high. Imagine frolicking through the three-and-a-half-inch-high grass adorned with little flowers, just like Julie Andrews. Who knows, maybe being able to sing in perfect pitch would come along with it. Yes, you might want to wear shoes, but that's not too much to ask.

If the idea of flowers growing in the grassy lawn just isn't quite achievable yet, well, there's always the golf course route. Take out some of that lawn and convert it into a home and dining hall for bees. It's all a matter of rethinking normal.

Citizen Science and the Great Sunflower Project

MY 'LEMON QUEEN' sunflowers seem to have recovered from their privations. I planted them in the big raised planter out by the street and have great hopes for them, despite their bad beginnings. When I found the sunflowers at a plant sale they were a bit wilted, and one had such a snaky stem that it felt the need to lean—or possibly lie down altogether. I bought these rather pitiful plants because I was afraid I wouldn't be able to find 'Lemon Queen' sunflowers anywhere else, since I'd found no seedlings last year. Theoretically, I could have started vast numbers from seed, but I'd left it a bit late. Plus, I don't have a way with seeds.

I wanted these 'Lemon Queen' seedlings so I could participate in the Great Sunflower Project, a nationwide citizen science project whose goal is to assess pollinators across the country, and 'Lemon Queen' sunflowers are the flagship plant for the project. By using the same cultivar throughout the country, the researchers can compare apples to apples or, in this case, sunflowers to sunflowers.

Citizen science and the bees

In 2007 the National Academy of Sciences published *Status of Pollinators in North America*. Among the many findings was the fact that we had no idea of the status of most of our native bees. For many native bees, there was little current or historical data on their ranges or abundance. Without some sort of baseline, it's impossible to know if a bee is doing well or doing poorly. For many bees we didn't even know where they nested or what they ate. We still don't.

When I think about what it would take to gather the amount of data needed to start making a baseline for our 4000 or so native bees, I just start shaking my head. Impossible. Years of bee collecting all over the country followed by an unimaginable number of bees needing identification. Buildings might collapse under the weight of the Schmitt boxes full of bees that would result from such a study.

Gretchen LeBuhn of San Francisco State University figured out a way around part of the problem: get ordinary people to collect data; we cost nothing. Of course, we the people have a very limited array of data we can collect about pollinators, given the scantiness of our knowledge base. Nevertheless, Gretchen came up with a plan that would let us help plug a hole or two in what we know about bees in the United States. She named this massive data collection effort the Great Sunflower Project.

The idea for creating this vast citizen science project came out of some work Gretchen had done with vineyard owners in Napa and Sonoma, California. She knew what bees had been lost from that area and wanted to find out what that meant for pollination. She realized that anyone could collect the data; they didn't actually need to be able to tell *Halictus* from *Andrena*. Gretchen went to the vineyard managers for help, and things worked so well that the next year she sent out emails to about forty people in the southeastern United States. Two months later she had 20,000 volunteers planning to count bees on sunflowers. It was 2008 and the Great Sunflower Project had officially begun.

The purpose of this project isn't to see if *Agapostemon virescens* or *Bombus sonorus* or any other individual species is doing well in a particular area. That's the kind of project that really does require expertise, an extraordinary number of hours, and lots of money. Instead, Gretchen has asked people to go out and count bees on one kind of plant, 'Lemon Queen' sunflowers, all across the United States. She could then use the data to get a sense for how many pollinators were visiting plants in different parts

A *Melissodes* female blending in with the sunflower, covered in pollen from antennae to ovipositor.

of the country and where pollination seemed to be going well—or poorly.

She's found some surprises in the data collected so far. Both California and the desert Southwest are rich in bee species, yet the data show surprisingly few bees visiting the sunflowers. She has theories about why—drought, lots of species of bees that are specialists on all sorts of plants that aren't sunflowers—but she doesn't know for sure. Nevertheless, thanks to ordinary people counting pollinators, some data are available that weren't in 2007 when the *Status of Pollinators in North America* came out. It's a place to start.

Bees and sunflowers

Sunflowers are North American native plants that belong to the Asteraceae family, once known as the Compositae. I like the old name because it tells you a lot about the flowers of this group. Members of this family are tricksters. What looks like one flower—a big yellow sunflower or a small yellow dandelion—is actually a group of small flowers, florets, clustered tightly together, making a composite. These florets fall into two groups.

The flowers of the outermost ring are called ray flowers and usually have one big petal pointing out. It's an odd duck of a flower if looked at individually, but when put together in a circle with all those petals poking out the whole yells "flower here." Inside the ring of ray flowers are the myriad little disc flowers all crammed together. A sunflower may have a thousand or more individual florets in each flower head, a convenient buffet for a bee.

When a bee lands on a sunflower, not every one of those thousand or more florets will be producing pollen and nectar. The florets start opening at the rim of the flower head and move inward. So a bee stopping by one day can mosey around and find a goodly number of flowers open and then stop back on that exact same sunflower the next day and find more food. Convenient. These little flowers are also easy to access when it comes to pollen and nectar collection, so it's no surprise that a lot of bees have a fondness for sunflowers.

A study done in Utah investigated how effective different kinds of bees are at pollinating sunflowers by looking at how many seeds resulted from a bee visit. Honey bees are often not great pollinators on an individual bee basis, and it's no surprise to find that they aren't wonder bees when it comes to sunflowers, averaging about two and a half seeds per visit. Bumble bees were worse, which is surprising. They didn't even manage to average one seed per visit. The big winner of the seed-making competition was *Melissodes agilis*, which averaged almost twenty-eight seeds per visit. Earlier research had given honey bees credit for the majority of sunflower pollination, but this study found that, at least where efficiency is concerned, honey bees are among the losers. Even more impressive were the different amounts of pollen the bees picked up. When they washed pollen off the bees (imagine the tiny bee bathtubs with itty bitty scrub brushes), honey bees left an average of 1778 pollen grains in the bath water. *Melissodes*—104,542.

A hundred thousand pollen grains—on one bee. I imagine the bee practically foundering under her load, flapping valiantly homeward to lay in these provisions for one of her babes. It's just another positive image to add to my collection for *Melissodes*, one of my favorite bees.

What's in a name?

People who name animals and plants often look to Latin or Greek for inspiration. Take *Melissodes*. *Mel* is Greek for "honey," and *melissa* is Greek for "honey bee." The suffix *-odes* means "like," so *Melissodes* is "honey bee–like." The species name *agilis* is an obvious one, "agile." So, *Melissodes agilis* is an agile, honey bee–like bee.

Sometimes the species part of the name may refer to a person. I found at least a dozen bees with the species name *rozeni*, including *Neolarra rozeni*. They are undoubtedly named after Jerry Rozen, cleptoparasite expert extraordinaire.

In scholarly papers the bee name sometimes has another name appended to it. This is the person who first identified that bee. Cockerell is a name that appears often. If Theodore Dru Alison Cockerell had been a bee, he would have been a generalist. He studied mollusks, bees and other insects, arachnids, fungi, mammals, fish, and plants. Those were the good old days when you didn't have to specialize. Robert Zuparko put together a list of all the organisms named by Cockerell and came up with 9043, ranging from the subspecies level all the way up to the superfamily.

Melissodes desponsa taking a break, using its mandibles to clamp onto a goldenrod stem.

Melissodes, long-horned bees

One of the nice things about *Melissodes* species, from the point of view of a bee admirer, is that they are big enough to see well with the naked eye. They tend to be stout bees and often have gray hair on their thorax and stripes on their abdomen. The females have a 1980s *Flashdance* feel going. The pollen-carrying hairs on their hind legs are abundant and feathery; they look like big old leg warmers. The males have very loooong antennae and an interesting way of sleeping.

Back in the early twentieth century, Phil and Nellie Rau wrote about the sleep of insects, including *Melissodes*. I always look to see if these two wrote anything about the bee I'm currently interested in. They describe bees with such joy, caring, and verve that I always find their work a pleasure to read.

The Raus found a group of *Melissodes obliqua* near St. Louis, Missouri, clustered on dead stems in a small burned patch of vegetation.

A *Melissodes robustior* female on *Dahlia coccinea*. Although called long-horned bees, *Melissodes* females don't have particularly long antennae. It is the males that give the genus their common name.

A *Melissodes robustior* male showing why members of this genus are called long-horned bees. ▲

Eucera dubitata is another member of the long-horned bee group. ◄

The surrounding live green plants held no bees. When they first walked through the burnt patch near dusk, the bees were settling in for the night but weren't fast asleep yet. Many of them got irked and flew up, buzzing about and refusing to light anywhere until the Raus left. They returned later, once the bees were good and asleep, and found twenty-eight bees sleeping on the burnt stems. Only three slept alone. The rest were settled, several together in cozy clusters, heads down and curled inward "as far as their chubby abdomens would permit."

The Raus returned repeatedly to the site. One day they marked some of the bees with white paint to see if the same bees were returning to the burnt patch each night. Regretfully, the experiment was foiled by a cow that came along during the day and destroyed the bees' sleeping spot. The experiment wasn't a complete failure, though, because the Raus found a few of the paint-splotched bees sleeping nearby and so assumed that at least some of those bees had called the burnt stems home. Another patch of plants had one old dead stem in the middle of a bunch of live green ones. The green stems were empty; the dead one covered with forty *Melissodes*.

When the Raus noted the gender of the *Melissodes* bees they observed, they were all males. Almost all of them preferred sleeping in clusters,

Another bee sleeping out in the cold, but without any buddies. This *Nomada* bee, a cleptoparasite, clasps onto a stem to sleep.

although one shunned his own kind and went off to sleep in the midst of a bunch of wasps.

Other *Melissodes* did things differently. The Raus found some sleeping flat out on top of sunflowers, one bee per flower head. Elsewhere they found three different species of *Melissodes* sleeping together like brothers, huddled on some dock plants. Years later Phil Rau wrote of two other *Melissodes* bees who had gone head first into some holes, but the holes were so small they could only get their heads in. Crazy ostrich bees.

Back in the garden with the 'Lemon Queen' sunflowers

When I got my sunflowers home I watered them, put them in the ground, and propped up the lazy one with sticks. A day or two later it fell over, and a snail munched on it. I went snail hunting. The raised bed that housed the sunflowers has a lip, so I knew where the nefarious grazers were likely to be hiding. Those I found got tossed out into the street. I feel only a little bad about it.

I propped the plant up again and kept an eye on it. All the sunflowers were growing well, seeming to have recovered from their early water stress and starting to form those big sunflower buds. My moment was coming. Then, one wilted. Not my strongest sunflower, but not my weakest either. This wasn't a little wilt but an "I'm pretty much dead" wilt. The soil was moist but not too moist. I didn't know what had happened. A few days later my best plant went. Boom. Wilting like the other for no obvious reason. Other plants in the bed were fine. This was bad. I was down to one sunflower with no backups.

A week later I saw that my last 'Lemon Queen' had fallen. I thought it had just slipped through the sticks propping it up, but no. The base was brown and woody. I had noticed this on the other ones but thought nothing of it. This sunflower had a gash in that woodiness, and ants crawled in and out. All my sunflowers were gone.

Robber bees

In the late 1970s, researchers at the University of California Davis watched bees on sunflowers; they were waiting for thieves. A rotund *Diadasia* female, a likely victim, had gathered pollen, stashing it away in the hairs on her hind legs. A honey bee came up behind her and used her mandibles and front legs to scrape some pollen off the *Diadasia*. The thief took her booty and groomed it onto herself. The scientists called all this thievery cleptolecty. The *Diadasia* put up with this heist, as did others of her genus. A victim might respond by lifting her leg, like she was trying to shake the honey bee off, but she didn't actually fly away until her pollen stores had been pillaged several times.

Most of the honey bees gathered pollen in the usual way—off the flower—but once one learned to steal, she was likely to be a repeat offender. Occasionally, a *Halictus* bee also stole from a *Diadasia*, but it seemed more accidental, as if she thought this well-laden leg was part of the flower. Periodically, the honey bees got so carried away they would bite the pollen pellets off another honey bee's leg.

Diadasia ochracea on bush mallow (*Malacothamnus*) with a pollen load worth stealing. She's also displaying remarkable flexibility with that leg.

Sunflowers and specialist bees

I thought of calling around to see if I could find more 'Lemon Queen' sunflowers, but the likely culprit for my wilt was *Sclerotina*, a soil-borne fungal pathogen with a nasty tendency to hang around for years. Planting more sunflowers in the same spot was pointless, and *Sclerotina* might well be resident elsewhere in the yard. I was disappointed. Some pretty cool things happen on sunflowers—the usual pollen and nectar gathering but also sometimes theft and sleep—all of which I'd like to see. Living in Seattle, I was unlikely to get the bees that are sunflower specialists, as conifer woodlands are not prime sunflower territory, but I had hoped for a few of their relatives.

When it comes to collecting pollen for the babes, some bees are pollen generalists and some bees are pollen specialists. Pollen generalists gather from whatever flower looks good at the moment. Often they go back to the same type of flower over and over because the pickings are good, and they know where the flowers are and how to get the groceries. A different bee of the same species might decide to devote herself to a completely different kind of flower. This flower constancy is good for pollination because an individual bee is moving pollen around among the same kind of plants, while the species as a whole visits a variety of flowers.

Pollen specialists are different. It's not a matter of individual choice, like it is for flower constancy. Specialists may be uber-specialists and use only one species of plant, or they may be only moderately picky and go after a group of related plants. Some of the bees that appear to like only one plant may actually find pollen acceptable from a variety of related plants. If, however, only one of their acceptable plants is in bloom when they fly, they appear to specialize on just that one plant.

I've known quite a few toddlers like these bees. Some toddlers want to eat nothing but pepperoni pizza: uber-specialists. Others might prefer pepperoni pizza but are fine with macaroni and cheese and possibly peanut butter and jelly sandwiches (as long as they are the proper brands). Woe upon the parents who try to feed one of these kids a hamburger, or some

A *Svastra obliqua* male in California showing the moderately long antennae of this genus. This species has a great fondness for sunflowers. ▲

There's no doubt about the gender of this *Svastra petulca*. Look at those pollen-laden legs! This bee is from Charleston, South Carolina. ◄

pear. (My son is a specialist. I made him try pear once—just once. He started choking so badly I had to turn him upside down.) Luckily, these specialist kids won't die if they have to eat something other than pepperoni pizza, but is the same true of specialist bees? What happens if a specialist bee gets the wrong pollen?

Not surprisingly, few studies have been done on feeding specialist bees the wrong pollen. Perhaps the surprising thing is that any studies have been done on it at all. It turns out that this kind of study is hard to do because many of the bees who are pollen specialists nest in the ground. Finding those larvae and digging them up in order to feed them alternate foods is difficult. It's much easier with bees that nest aboveground, where you can put out blocks or stems as trap nests and then bring them into the lab.

The few studies that have been done suggest that bees don't generally specialize because the larvae *must* have their special pollen. So why do bees evolve into specialists if many of them probably won't keel over from the wrong pollen?

A group of bees may become pollen specialists for several reasons. Some bees pick the same plant they were fed as larvae, even if it is the wrong food for the kind of bee they are. They pick what is familiar: comfort food. Other bees, even when fed the wrong food as larvae, still pick the food that their kind of bee is supposed to eat, and so the food choice for them seems to be truly inherited. Some bees might specialize on a plant that tends to be out in abundance when they wake. Others may pick a plant for which they have some trait that makes them particularly able gatherers, such as hooked hairs for getting pollen out of difficult flowers or especially sensitive eyes for foraging at dawn or dusk.

Maybe I should stop expecting all bees to behave the same way just because we've placed them all in a category that we call "bee." The world has 20,000 species of bees spread over nine families. To toss them all together and assume that they will look the same or act the same or evolve in the same way is a bit like expecting a grizzly bear to be like a seal. I didn't pull those two animals out of nowhere. Taxonomically speaking, a grizzly bear and a seal are as closely related as two bee species from different families. We don't expect the bear and the seal to eat the same things or bed down in the same way. Why then do we expect it of bees?

Diadasia: bees that build

When I was investigating sunflower bees, I ran across *Diadasia enavata*, a fat fine-looking bee that lives in the western United States. Its common name is the sunflower chimney bee. Since I already knew about its fondness for sunflowers, that part of its common name didn't surprise me, but I did wonder where the chimney bit came from—right up until I saw a photo of a nest site for this bee. It sent me off on a hunt into the world of *Diadasia* nest building.

Some *Diadasia* species build chimneys or turrets at their nest entrances. Which is the better name for these constructions rather depends on the purpose of that structure, and I've found no one who seems to know for sure why these bees build towering nest entrances. I'm going to call them turrets because I like the idea that every bee's nest is her castle.

Some researchers watched *Diadasia afflicta* bees in Austin, Texas, as they set up their nests. The bees picked open areas in hard-packed ground to start their nests. A female would fly low over a nest site, periodically stopping to walk around and tap the soil with her antennae. After she found a good spot, she'd start digging, using her mandibles and forelegs. Early in the digging process, the bee would put her head on the ground and rotate in circles. Was she trying to drill with her head? Sometimes, the bee would fly off in the midst of digging, coming back to regurgitate a fluid, probably nectar, which she then mixed into the dirt. Anything to make the digging easier because getting through that top layer of soil could clearly be difficult. A bee might spend up to forty-five minutes making the beginning of a hole. Calling this initial excavation a "hole" is truly giving it delusions of depth, since it only ran to about four millimeters, or a bit more than an eighth of an inch.

Once the bee had gotten started, the next step went quickly with the female digging down and shoving the excavated material out of the hole

Ho ho ho! *Diadasia enavata*, the sunflower chimney bee, heading home.

to make a little mound (called a tumulus). Eventually, she stopped digging and came up to work on her turret. The bee would start head down in the burrow with just the tip of her abdomen poking out. She'd rotate around the edge of the opening, pushing up moistened bits of soil and tapping them into place with her abdomen, eventually making a smooth round opening. After a while, she would come out and fly low to fan away the loose bits of dirt, leaving her masterpiece exposed—a turret cemented with regurgitated nectar.

Sometimes, in the night, bits of pollen were added to the top of a turret. These pollen grains were missing some of their nutritious inner parts, suggesting that they had gone through something's gut, a theory confirmed by the substance the pollen was embedded in—bee poop. So the bee had a meal, pooped, cleaned house, and decided that the cleanings made nice masonry material? A hard day of foraging and you come home to house cleaning; it's every working mother's lot.

The Great Sunflower Project, phase 2

I had known when I bought my sunflowers for the Great Sunflower Project that I wasn't going to have a *Diadasia* building turrets in my yard since none live in Seattle. I had hoped to see a *Melissodes* or two on my sunflowers, but clearly that moment has passed. I still want to help in bee research though. Fortunately, I can still do that since the Great Sunflower Project has another project, one where you don't need sunflowers to participate. Pick any plant you want, as long as you can put a name on it, and count the pollinators visiting it. This I can do, although it takes me a while.

In the summer of 2016, Gretchen LeBuhn told me that 100,000 people have signed up for the Great Sunflower Project, but only about five percent participate. I get it. I always seem to think about counting bees when it's raining and no bees are out or when I'm walking out the door to do something else. Finally, a month or so after my great sunflower death, good bee weather and opportunity arrive together. It's time to count.

I have two excellent bee plants right out front to choose from. My neighbor has a hebe that is a bee magnet in June, and I have several *Salvia* 'Indigo Spires' that are always full of bumble bees. Either would be a great choice, or so I thought.

My neighbor's hebe is a pleasure for a bee watcher. I've seen *Anthidium manicatum* (doing its thing: patrolling and having sex), three species of bumble bees (not being nearly as interesting as *Anthidium*), honey bees (also merely gathering), *Agapostemon* (which didn't need to be interesting, as its beauty alone was enough), and a host of small, zippy, black insects that I thought might be halictids but turned out to be some kind of wasp.

The Great Sunflower Project has rules, naturally, for how you count pollinators. For plants whose flowers grow on spikes like those of the salvia and the hebe, you're supposed to count the number of flowers on the spike that you're watching. Each spike on that hebe has a lot of flowers. I'd probably have to pick the spike after the count and do a he-loves-me, he-loves-me-not thing to get an accurate number. Certainly doable. My neighbor won't miss one spike because the real problem with using this plant is that it has a zillion spikes. The likelihood of anything landing on the spike that I've chosen to watch in a time span I can tolerate (ten to fifteen minutes tops) is unlikely. Also, I don't know which cultivar the hebe is.

I take the easier route and turn to my *Salvia* 'Indigo Spires'. It has fewer spikes than the hebe, and I realize that I can pick a spike with a bee already on it so I don't come up with zero bees. Is that cheating? It seems wrong to come up with zero bees for a plant that clearly has lots. Right or not, that's what I do. I take my little notepad and start my timer. I watch my spike of eleven flowers for seven minutes. I count four honey bees and one bumble bee. I do another spike and get six honey bees. I enter the data. I am now part of the five percent.

One of the cool things about the Great Sunflower Project website is that it tracks your data for you. The average number of pollinators per hour per flower for my yard (well, one plant in my yard) is around forty-seven, which handily beats the regional average of almost twelve. I knew the salvia was a good bee plant, but I have to say seeing that regional average makes me feel a bit competitive. I don't want to see my numbers drop, which should work well for the bees.

My neighbor's hebe, flower spikes galore, with two *Anthidium manicatum* (wool carder bees) mating.

The purpose of this Great Sunflower Project program is different from the one with the 'Lemon Queen' sunflowers. Over time, Gretchen will be able to develop pollinator plant lists that are truly local, often right down to the cultivar. A real need exists for such lists.

In search of lists

The web has plenty of pollinator plant lists. Some are generic and seem meant to cover the entire country. Others are regional but there may be a surprising amount of overlap between regions, which makes me wonder how much science is behind them. Most of these lists leave me feeling dissatisfied. One regional list I found had "asters" down as a pollinator plant. Holy smoke, how many kinds of asters are there? Dozens? Hundreds? How on earth should I figure out which one to pick? Most plants on this particular list were native. Perhaps they meant a native aster? Bah! I want more.

The conventional wisdom is to use native plants for native animals because they coevolved together. Certainly some of the insects that are herbivores are very limited in what they eat. Look at how persnickety moth and butterfly larvae can be toward their food. Many bees, it turns out, aren't that picky. For those of us who are happy with some native plants but also like our ornamental exotics, this is a relief. Gordon Frankie told me once that there are "lots of non-natives that do wonderful things for bees." Gordon is totally focused on which plants bees like, whatever their country or region of origin may be. He loves Provence lavender (*Lavandula ×intermedia* 'Provence') because his research group has found nearly fifty bee species on it in California.

I really do have a bit of a thing about these lists. I want specifics, which is a little unreasonable of me. Doing the trials and research to put together science-based lists takes time and money. Some folks at Penn State (and affiliated Master Gardeners) did side-by-side trials of species and cultivars. It took years. Surprisingly, they found that the straight species wasn't always the pollinators' favorite. I'd assumed it would be because plant breeders certainly aren't selecting for what a bee wants. The study also showed that sometimes a plant that looks good initially may not hold up well over time. *Coreopsis verticillata* 'Zagreb' brought in more pollinators than the straight species when the Penn State folks started the trials, but after several years 'Zagreb' started to get spotty while the species looked fabulous. So which plant was helping pollinators the best?

Others share my frustration with the dearth of good lists. Bernie Mach, a graduate student at the University of Kentucky, told me, "You see all these lists. The best plants to plant for bees. Plant these! You'll have so many bees." Often, she says, no scientific data support these lists. She and her advisor, Dan Potter, are looking at the pollinator potential of an often-neglected group of plants: those that are woody.

While we are talking about the project, Dan pulls Schmitt boxes down off the shelves. All the boxes are arranged by plant species. The *Philadelphus* (mock orange) box has loads of bees but almost all of them are the *Philadelphus* pollen specialist *Chelostoma philadelphi*. Some plants have lots of honey bee visitors, whereas others have a slew of different types of bees visiting. Bernie and Dan's ultimate goal is to put out a science-based list of garden-worthy trees and shrubs that will supply nectar and pollen to pollinators throughout the season. This project, like the one at Penn State, will take years to complete.

Given the time and money needed to generate science-based pollinator plant lists, many regions won't have any useful ones. Fortunately, this is a place where ordinary people have the power to accelerate the acquisition of knowledge. Most of us may not have the power of great knowledge of bees, but we do have the power of great numbers. Go forth to your yard, your neighbor's yard, or a park and find a plant with pollinators on it and count them. Then go straight home and immediately enter the data at the Great Sunflower Project website. Delay is deadly and likely leads to data never getting entered. The data that people collect will be sifted, analyzed, and compiled into lists—regional, detailed, and no doubt delightful. We may not know which bees are on the plants, but we will certainly know which plants the bees like best. Plus, it makes you go outside. Your

Chelostoma philadelphi

Chelostoma seems to be a genus filled with nondescript dark bees; *Chelostoma philadelphi* certainly fits that description. They are awfully hard workers, though. Some were found nesting in old beetle burrows in a fig tree. The female would pick a hole and clean out all the nasty beetle detritus (mostly beetle poop). Then she'd go collect provisions, lay her egg, and put up a door made of mud and pebbles that might take her twenty trips and over two hours to complete.

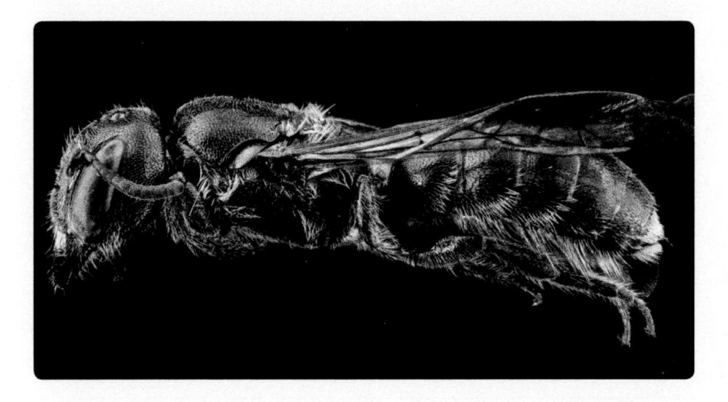

Chelostoma philadelphi won't win any beauty awards.

neighbors will have a reason to start up a conversation when they see you standing in your jammies some morning, staring at one of their plants because you noticed some really interesting early-rising bees on it.

Ten or fifteen minutes a week of observation done by thousands of people starts to add up to some real data. We may never get a comprehensive baseline on the status of all of North America's bees, but regular people can contribute to our knowledge base on bees and at the same time we provide those bees with food to eat. How cool is that?

The Power of Bees

BEES CAN INSPIRE.

Gail Langellotto-Rhodaback is the statewide Master Gardener program coordinator for Oregon and a professor at Oregon State University. Before moving to Oregon she worked in New York City, where she conducted a survey of bees in that most urban of cities. One day she was out collecting water pan traps (bee bowls) next to a high school right when school was letting out for the day. Some of the kids came over to see what was going on.

"We were all standing in a circle," Gail says, "looking at what was in the bowls and we [the scientists] were pointing out the different insects that we'd collected in the water pan trap. And it was really kind of cool because they didn't know that there was such a thing as a good insect."

At which point the cops rolled up and hit their siren. With everyone standing around in a circle, "I guess we looked like we were doing a drug deal," Gail says.

The cops walked up to see what was happening, and then they wanted to know about the insects in the bee bowls too. "It was really, like, a beautiful moment," Gail said. "In East Harlem, we had teenagers and scientists and graduate students and cops all looking at these bees in water pan traps, interested and asking questions."

Bees have power.

They have the obvious power of pollination and supplying us with many of our favorite foods. They also have an unexpected superpower—the ability to form connections and build community among people. That

This little cutie, *Anthophora curta*, was caught in Arizona by one of the people I met at the Bee Biodiversity Initiative and photographed at Sam Droege's U.S. Geological Survey bee lab. Below many of the bee photos on the lab's website someone has added these lines from John Keats's poem "Ode on a Grecian Urn": "Beauty is truth, truth beauty—that is all / Ye know on earth and all ye need to know."

moment in New York City is just one example. People come together to volunteer at bee labs or help with bee surveys. Some use vacation time to take bee classes and hunt for bees. I have been astounded, again and again, by the bee people I've met. They've shared their time, their knowledge, and their passion with a complete stranger who sends them an email, asking to talk or come spend a day in their life.

Bees are resilient.

We may think the world is falling apart and an individual can do little to help stop it. That is not true for bees. If we just stop kicking the bees quite so hard, we can help them—and see the results almost immediately. Renounce pesticides. Plant flowers that bees in your area like. Be a little slovenly in the garden; leave some old broken stems and let a little bare dirt show. The bees will come.

Bees are diverse.

It's probably this vast diversity that has struck me the most in all the research that I've done on bees. Most people think of honey bees when they hear the word *bee* or, even worse, they envision a yellow jacket or some other kind of wasp. Twenty thousand species rife with differences being reduced to either a very unusual outlier of the group or something that is not a member of the group at all.

When you think of bees, think instead of those males guarding their territories, sleeping together, hanging out at the bee bar waiting for females, and getting into battles and biting each other on the legs. Think of the females that excavate deep mines to make homes for their babes or build turret entrances to their nests out of dirt and regurgitated nectar, with bee poop crenellations. Think of the little cleptoparasite larvae taking down their gigantic foes and the Cinderella *Ceratina* bees out gathering for their siblings with no hope of a future home of their own. These are our bees.

Bees have changed my life.

This book started when I overcame my introvert's nature and contacted Robbin Thorp to ask if I could join him on a Franklin's bumble bee hunt. Roaming those slopes with Robbin and listening to him talk made me think there were indeed bee tales to be told—and I wanted to tell them—even if the prospect of approaching people to ask for interviews scared the willies out of me. I never thought I'd write a book, let alone feel like I've found my calling, and it's all thanks to bees.

That trip to meet Robbin was in August 2014. I haven't been able to make it back to Mount Ashland for another of his bee hunts in the intervening years. I plan to go again, hopefully this coming year. I'll drive down I-5 to that little oval of land where the Franklin's bumble bee once flew. I'll meet Robbin and any of the other bee enthusiasts and experts who join him, and we'll hunt. We won't put out bee bowls. Bumble bees tend to be good at escaping from them. And, holy smoke, what if we drowned a Franklin's? I'm sure no cops will show up to ask what we're doing, but any hikers who are around will no doubt ask what we're up to and go away knowing a little something about the world that they hadn't before. It's part of the bees' superpower. And, who knows, maybe this will be the year that black-bottom bee returns.

Metric Conversions

inches	cm
¼	0.6
½	1.3
1	2.5
2	5.1
3	7.6
4	10
5	13
6	15
7	18
8	20
9	23
10	25

feet	m
1	0.3
2	0.6
3	0.9
4	1.2
5	1.5
6	1.8
7	2.1
8	2.4
9	2.7
10	3
100	30
1000	300

miles	km
1	1.6
2	3.2
3	4.8
4	6.4
5	8.0
10	16
20	32
30	48
100	160

Selected References

What a Bee Is

Cole, Adam. Honey, it's electric: bees sense charge on flowers. npr.org/2013/02/22/172611866/honey-its-electric-bees-sense-charge-on-flowers.

Crepet, William, and Karl Niklas. 2009. Darwin's second "abominable mystery": Why are there so many angiosperm species? *American Journal of Botany* 96(1): 366–381.

Engel, Michael S., Ismael A. Hinojosa-Diaz, and Alexandr P. Rasnitsyn. 2009. A honey bee from the Miocene of Nevada and the biogeography of *Apis* (Hymenoptera: Apidae: Apini). *Proceedings of the California Academy of Sciences* 4(60): 23–38.

Frankie, Gordon, Robbin Thorp, Rollin Coville, and Barbara Ertter. 2014. *California Bees and Blooms: A Guide for Gardeners and Naturalists*. Berkeley, CA: Heyday.

National Honey Bee Health Stakeholder Conference Steering Committee. 2012. *Report on the National Stakeholders Conference on Honey Bee Health*. Washington, DC: U.S. Department of Agriculture.

Ollerton, Jeff, Rachel Winfree, and Sam Tarrant. 2011. How many flowering plants are pollinated by animals? *Oikos* 120(3): 321–326.

Pollinator Partnership. Primer on pollination and pollinators. pollinator.org/Resources/facts.Primer.pdf.

Ramanujan, Krishna. Insect pollinators contribute $29 billion to U.S. farm income. news.cornell.edu/stories/2012/05/insect-pollinators-contribute-29b-us-farm-income.

Rosner, Hillary S. 2013. Return of the natives: how wild bees will save our agricultural system. *Scientific American* September 1. Available via scientificamerican.com/article/return-of-the-natives-how-wild-bees-will-save-our-agricultural-system.

Smithsonian Institution. Numbers of insects (species and individuals). si.edu/Encyclopedia_SI/nmnh/buginfo/bugnos.htm.

University of Zurich. 2013. New fossils push the origin of flowering plants back by 100 million years to the Early Triassic. *Science Daily* October 1. Available via sciencedaily.com/releases/2013/10/131001191811.htm.

Vaknin, Y., S. Gan-Mor, A. Bechar, B. Ronen, and D. Eisikowitch. 2000. The role of electrostatic forces in pollination. *Plant Systematics and Evolution* 222(1/4): 133–142.

Xerces Society for Invertebrate Conservation. Native bee pollination of cherry tomatoes. xerces.org/wp-content/uploads/2008/10/factsheet_cherry_tomato_pollination.pdf.

A Bee for All Seasons: *Apis mellifera*, the European Honey Bee

Almond Board of California. 2014. Almond Almanac: Annual report 2014. almonds.com/sites/default/files/content/attachments/2015_almond_almanac_annual_report.pdf.

Almond Board of California. 2015. California almond industry facts. almonds.com/pdfs/california-almond-industry-facts.pdf.

American Bee Journal. The first human uses of beeswax have been established in Anatolia in 7000 BCE. americanbeejournal.com/first-human-uses-beeswax-established-anatolia-7000-bce/.

Berenbaum, May, Peter Bernhardt, Stephen Buchmann, et al. 2007. *Status of Pollinators in North America*. Washington, DC: National Academies Press.

Buchmann, Stephen L., and Banning Repplier. 2005. *Letters from the Hive: An Intimate History of Bees, Honey, and Humankind*. New York: Bantam.

California Department of Food and Agriculture. The 2015 California almond acreage report. nass.usda.gov/Statistics_by_State/California/Publications/Fruits_and_Nuts/2016/201605almac.pdf.

Desneux, Nicolas, Axel Decourtye, and Jean-Marie Delpuech. 2007. The sublethal effects of pesticides on beneficial arthropods. *Annual Review of Entomology* 52: 81–106.

Doll, David. 2013. IGRs at bloom: Bad idea? thealmonddoctor.com/2013/01/17/igrs-at-bloom-bad-idea.

Engel, Michael S., Ismael A. Hinojosa-Diaz, and Alexandr P. Rasnitsyn. 2009. A honey bee from the Miocene of Nevada and the biogeography of *Apis* (Hymenoptera: Apidae: Apini). *Proceedings of the California Academy of Sciences* 4(60): 23–38.

Honey Bee Suite. How much honey do bees need for winter? honeybeesuite.com/how-much-honey-should-i-leave-in-my-hive.

Ingraham, Christopher. 2015. Call off the bee-pocalypse: U.S. honeybee colonies hit a 20-year high. *The Washington Post*, July 23. Available via washingtonpost.com/news/wonk/wp/2015/07/23/call-off-the-bee-pocalypse-u-s-honeybee-colonies-hit-a-20-year-high.

Koeniger, N., G. Koeniger, and S. Wongsiri. 1989. Mating and sperm transfer in *Apis florea*. *Apidologie* 20(5): 413–418.

Langstroth, L. L. 1853. *Langstroth on the Hive and the Honey-Bee: A Bee Keeper's Manual*. Medina, OH: A. I. Root Company.

Life Cycle of the Honey Bee. *Get Buzzing about Bees*. buzzaboutbees.net/honey-bee-life-cycle.html.

Michreit, Kathrin, Haike Ruhnke, Jakob Wegener, and Kaspar Bienefeld. 2016. Effects of an insect growth regulator and a solvent on honeybee (*Apis mellifera* L.) brood development and queen viability. *Ecotoxicology* 25(3). Available via ncbi.nlm.nih.gov/pubmed/26821233.

Moore, Philip A., Michael E. Wilson, and John A. Skinner. Honey bee queens: evaluating the most important colony member. articles.extension.org/pages/73133/honey-bee-queens:-evaluating-the-most-important-colony-member.

Mullin, Christopher A., Maryann Frazier, James L. Frazier, Sara Ashcroft, Roger Simonds, Dennis van Engelsdorp, and Jeffery S. Pettis. 2010. High levels of miticides and agrochemicals in North American apiaries: Implications for honey bee health. *PLoS One*, March 19. Available via http://journals.plos.org/plosone/article?id = 10.1371/journal.pone.0009754.

Mussen, Eric. Apiculture Newsletter archive. entomology.ucdavis.edu/Faculty/Eric_C_Mussen/Apiculture_Newsletter.

National Honey Bee Health Stakeholder Conference Steering Committee. 2012.

Report on the National Stakeholders Conference on Honey Bee Health. Washington, DC: U.S. Department of Agriculture.

Oertel, E., E. C. Martin, and N. P. Nye. 1980. *Beekeeping in the United States.* Agricultural Handbook no. 335. Washington, DC: U.S. Department of Agriculture.

Pearson, Gwen. Women work to save native bees of Mexico. wired.com/2014/03/women-work-save-native-bees-mexico.

Ramanujan, Krishna. Insect pollinators contribute $29 billion to U.S. farm income. news.cornell.edu/stories/2012/05/insect-pollinators-contribute-29b-us-farm-income.

Reyes-Gonzalez, Alejandro, Andres Camou-Guerrero, Octavio Reyes-Salas, Arturo Argueta, and Alejandro Casas. 2014. Diversity, local knowledge and use of stingless bees (Apidae: Meliponini) in the municipality of Nocupetaro, Michoacan, Mexico. *Journal of Ethnobiology and Ethnomedicine* 10: 47.

Rucker, Randal R., Walter N. Thurman, and Michael Burnett. 2001. An empirical analysis of honeybee pollination markets. Paper presented at the American Agricultural Economics Association, Chicago, Illinois, May 14.

Seeley, Thomas D. 2010. *Honeybee Democracy.* Princeton, NJ: Princeton University Press.

Villanueva-G, Rogel, David W. Roubik, and Wilberto Colli-Ucan. 2005 Extinction of *Melipona beecheii* and traditional beekeeping in the Yucatan Peninsula. *Bee World* 86(2): 35–41.

Watanabe, Myrna E. 2008. Colony collapse disorder: many suspects, no smoking gun. *BioScience* 58(5): 384–388.

White House Office of the Press Secretary. Presidential Memorandum: Creating a federal strategy to promote the health of honey bees and other pollinators. whitehouse.gov/the-press-office/2014/06/20/presidential-memorandum-creating-federal-strategy-promote-health-honey-b.

Did Greenhouse Tomatoes Kill the Last Franklin's Bumble Bee?

APHIS. *Animal Plant Health Inspection Service Plant Protection and Quarantine Plant Health Programs Briefing Paper on Importation and Interstate Transport of Bumble Bees.* March 17, 2009.

Bailey, L. H. 1891. Experiments in the forcing of tomatoes. *Bulletin of the Agricultural Experiment Station, Cornell University, Department of Agriculture* 1: 45–61. Available via hdl.handle.net/2027/uiug.30112019743498?url append = %3Bseq = 9.

Buchmann, Stephen L., and James P. Hurley. 1978. A biophysical model for buzz pollination in angiosperms. *Journal of Theoretical Biology* 72(4): 639–657.

Cameron, Sydney A., Jeffrey D. Lozier, James P. Strange, Jonathan B. Koch, Nils Cordes, Leellen F. Solter, and Terry L. Griswold. 2011. Patterns of widespread decline in North American bumble bees. *Proceedings of the National Academy of Sciences* 108(2): 662–667.

Cameron, Sydney A., Haw Chuan Lim, Jeffrey D. Lozier, Michelle A. Duennes, and Robbin Thorp. 2016. Test of the invasive pathogen hypothesis of bumble bee decline in North America. *Proceedings of the National Academy of Sciences USA* 113(16): 4386–4391.

Colla, Sheila R., Michael C. Otterstatter, Robert J. Gegear, and James D. Thomson. 2006. Plight of the bumble bee: Pathogen spillover from commercial to wild populations. *Biological Conservation* 129(4): 461–467.

Epsy, Sam. 2014. Letter from Sec. of Agriculture Mike Espy to Congressman Sam Farr Re Bumble Bees. June 1994.

Flanders, W. V., Wayne Wehling, and A. L. Craghead. 2003. Laws and regulations on the import, movement, and release of bees in the United States. In Karen Strickler and James Harley Cane, eds. *For Nonnative Crops, Whence Pollinators of the Future?* Annapolis, MD: Entomological Society of America.

Goulson, Dave. 2014. *A Sting in the Tale: My Adventures with Bumblebees.* London: Picador.

Michener, Charles D. 1962. An interesting method of pollen collecting by bees from flowers with tubular anthers. *Revista de Biología Tropical* 10(2): 167–175. Available via ots.ac.cr/rbt/attachments/volumes/vol10-2/06-Michener-Pollen.pdf.

Murray, Tomás E., Mary F. Coffey, Eamonn Kehoe, and Finbarr G. Horgan. 2013. Pathogen prevalence in commercially reared bumble bees and evidence of spillover in conspecific populations. *Biological Conservation* 159: 269–276.

Sladen, Frederic. 1912. *The Humble-Bee: Its Life-History and How to Domesticate It, with Descriptions of All the British Species of* Bombus *and* Psithyrus. London: Macmillan and Co.

Szabo, Nora D., Sheila R. Colla, David L. Wagner, Lawrence F. Gall, and Jeremy T. Kerr. 2012. Do pathogen spillover, pesticide use, or habitat loss explain recent North American bumblebee declines? *Conservation Letters* 5(3): 232–239.

Thorp, Robbin, Sarina Jepson, Sarah Foltz Jordan, Elaine Evans, and Scott Hoffman Black. 2010. Petition to list Franklin's bumble bee *Bombus franklini* (Frison), 1921 as an endangered species under the Endangered Species Act. xerces.org/wp-content/uploads/2010/06/bombus-franklini-petition.pdf.

Velthuis, Hayo H. W., and Adriaan van Doorn. 2006. A century of advances in bumblebee domestication and the economic and environmental aspects of its commercialization for pollination. *Apidologie* 37(4): 421–451.

Welch, Craig. 2000. A brief history of the spotted-owl controversy. *The Seattle Times*, August 6. Available via community.seattletimes.nwsource.com/archive/?date = 20000806&slug = 4035697.

Williams, Paul, Robbin Thorp, Leif Richardson, and Sheila Colla. 2014. *Bumble Bees of North America: An Identification Guide*. Princeton, NJ: Princeton University Press.

Winter, Kimberly, Laurie Adams, Robbin Thorp, David Inouye, Liz Day, John Ascher, and Stephen Buchmann. 2006. *Importation of Non-Native Bumble Bees into North America: Potential Consequences of Using* Bombus terrestris *and Other Non-Native Bumble Bees for Greenhouse Crop Pollination in Canada, Mexico and the United States*. A White Paper of the North American Pollinator Protection Campaign. Available via pollinator.org/Resources/BEEIMPORTATION_AUG2006.pdf.

Xerces Society for Invertebrate Conservation. 2010. *Petition before the United States Department of Agriculture Animal and Plant Health Inspection Service: Petition Seeking Regulation of Bumble Bee Movement*. Available via xerces.org/petition/xerces-bumblebee-petition-to-aphis.pdf.

Osmia lignaria, the Great and Glorious BOB

Bosch, Jordi, and William Kemp. 1999. Exceptional cherry production in an orchard pollinated with blue orchard bees. *Bee World* 80(4): 163–173.

Bosch, Jordi, and William Kemp. 2001. *How to Manage the Blue Orchard Bee as an*

Orchard Pollinator. Beltsville, MD: Sustainable Agriculture Network.

Bosch, Jordi, William Kemp, and Glen E. Trostle. 2006. Bee population returns and cherry yields in an orchard pollinated with *Osmia lignaria* (Hymenoptera: Megachilidae). *Journal of Economic Entomology* 99(2): 408–413.

Cane, James H., and V. J. Tepedino. 2016. Gauging the effect of honey bee pollen collection on native bee communities. *Conservation Letters* May: 1–6.

Cane, James H., Terry L. Griswold, and F. D. Parker. 2007. Substrates and materials used for nesting by North American *Osmia* bees (Hymenoptera: Apiformes: Megachilidae). *Annals of the Entomological Society of America* 100(3): 350–358.

Frankie, Gordon, Robbin Thorp, Rollin Coville, and Barbara Ertter. 2014. *California Bees and Blooms: A Guide for Gardeners and Naturalists*. Berkeley, CA: Heyday.

Griffin, Brian L., and Sharon Smith. 1999. *The Orchard Mason Bee: The Life History, Biology, Propagation, and Use of a North American Native Bee*. 2nd ed. Bellingham, WA: Knox Cellar Publishing Co.

National Agricultural Statistics Service. 2011. *Washington Tree Fruit Acreage Report 2011*. Washington, DC: U.S. Department of Agriculture.

Parker, F. D., and V. J. Tepedino. 1982. A nest and pollen-collection records of *Osmia sculleni* Sandhouse, a bee with hooked hairs on the mouthparts (Hymenoptera: Megachilidae). *Journal of the Kansas Entomological Society* 55(2): 329–334.

Pitts-Singer, Theresa L., and James H. Cane. 2011. The alfalfa leafcutting bee, *Megachile rotundata*: The world's most intensively managed solitary bee. *Annual Review of Entomology* 56(1): 221–237.

Torchio, Philip. 1984. The nesting biology of *Hylaeus bisinuatus* Forster and development of its immature forms (Hymenoptera: Colletidae). *Journal of the Kansas Entomological Society* 57(2): 276–297.

Wilson, Joseph S., and Olivia Messinger Carril. 2015. *The Bees in Your Backyard: A Guide to North America's Bees*. Princeton, NJ: Princeton University Press.

Bees, Blueberries, Budworms, and Pesticides

Argall, John. Native bees that pollinate wild blueberries. New Brunswick Department of Agriculture and Rural Development. www2.gnb.ca/content/gnb/en/departments/10/agriculture/content/bees/native_bees.html.

Batra, S. W. T. 1980. Ecology, behavior, pheromones, parasites and management of the sympatric vernal bees *Colletes inaequalis*, *C. thoracicus* and *C. validus*. *Journal of the Kansas Entomological Society* 53(3): 509–538.

Boulanger, L. W., G. W. Wood, E. A. Osgood, and C. O. Dirks. 1967. *Native Bees Associated with Low-Bush Blueberry in Maine and Eastern Canada*. Maine Agricultural and Forest Experiment Station Technical Bulletin TB-26. Available via digitalcommons.library.umaine.edu/cgi/viewcontent.cgi?article=1165&context=aes_techbulletin.

Bushmann, Sara, and Francis Drummond. 2015. Abundance and diversity of wild bees (Hymenoptera: Apoidea) found in lowbush blueberry growing regions of Downeast Maine. *Environmental Entomology* 44(4): 975–989.

Canadian Press. 1994. N. B. suspends air war against budworm. *Hamilton Spectator (Ontario, Canada)*, February 18, sec. Metro.

Canadian Press. 1995. Fenitrothion aerial spraying to be phased out over 4 years. *The Globe and Mail (Canada)*, April 19.

Casey, Quentin. 2012. Budworm: Economics and ecology. *New Brunswick Telegraph Journal*. January 3. Available via niche-canada.org/wp-content/uploads/2014/05/Telegraph-Journal-Article.pdf.

Crandall, Esther. 1976. Province budworm spraying done; foes take second. *Bangor Daily News*, July 7.

Gill, Richard J., Oscar Ramos-Rodriguez, and Nigel E. Raine. 2012. Combined pesticide exposure severely affects individual- and colony-level traits in bees. *Nature* 491(7422): 105–108.

Heinrich, Bernd. 1979. "Majoring" and "minoring" by foraging bumblebees, *Bombus vagans*: An experimental analysis. *Ecology* 60(2): 245–255.

Holyoke, John. 2016. It's coming: Maine readies for new battle with spruce budworm. *Bangor Daily News*, August 25. Available via bangordailynews.com/2015/12/10/outdoors/its-coming-maine-readies-for-new-battle-with-spruce-budworm.

Jarman, Walter M., and Karlheinz Ballschmiter. 2012. From coal to DDT: The history of the development of the pesticide DDT from synthetic dyes till *Silent Spring*. *Endeavour* 36(4): 131–142.

Javorek, S. K., K. E. Mackenzie, and S. P. Vander Kloet. 2002. Comparative pollination effectiveness among bees (Hymenoptera: Apoidea) on lowbush blueberry (Ericaceae: *Vaccinium angustifolium*). *Annals of the Entomological Society of America* 95(3): 345–351.

Kevan, Peter. 1975. Forest application of the insecticide fenitrothion and its effect on wild bee pollinators (Hymenopter: Apoidea) of lowbush blueberries (*Vaccinium* spp.) in southern New Brunswick, Canada. *Biological Conservation* 7(4): 301–309.

Kevan, Peter G. 1977. Blueberry crops in Nova Scotia and New Brunswick: Pesticides and crop reductions. *Canadian Journal of Agricultural Economics* 25(1): 61–64.

Kevan, Peter, and M. Collins. 1974. Bees, blueberries, birds and budworms. *The Osprey* V: 54–62.

Kucera, Daniel R., and Peter W. Orr. *Spruce Budworm in the Eastern United States*. Forest Insect and Disease Leaflet no. 160. U.S. Department of Agriculture Forest Service. Available via na.fs.fed.us/spfo/pubs/fidls/sbw/budworm.htm.

Leonard Lab. Buzz pollination. anneleonard.com/buzz-pollination.

McLaughlin, Mark. 2011. Green shoots: Aerial insecticide spraying and the growth of environmental consciousness in New Brunswick, 1952–1973. *Acadiensis* 40(1): 3–23.

Müller, Paul H. 1948. Dichloro-diphenyl-trichloroethane and newer insecticides. Nobel lecture. nobelprize.org/nobel_prizes/medicine/laureates/1948/muller-lecture.pdf.

National Site for the Regional IPM Centers. OPP Pesticide Ecotoxicity Database. Pesticide: DDT. ipmcenters.org/ecotox/Details.cfm?RecordID = 11281.

Osgood, E. A. 1989. Biology of *Andrena crataegi* Robertson (Hymenoptera: Andrenidae), a communally nesting bee. *Journal of the New York Entomological Society* 97(1): 56–64.

Plowright, R. C., B. A. Pendrel, and I. A. McLaren. 1978. The impact of aerial fenitrothion spraying upon the population biology of bumble bees (*Bombus* Latr.: Hym.) in south-western New Brunswick. *The Canadian Entomologist* 110(11): 1145–1156.

Rau, Phil, and Nellie Rau. 1916. Notes on the behavior of certain solitary bees. *Journal of Animal Behavior* 6(5): 367–370.

Sandberg, L. Anders, and Peter Clancy. 2002. Politics, science and the spruce budworm in New Brunswick and Nova Scotia. *Journal of Canadian Studies* 37(2): 164.

Schrader, Martha Northam, and Wallace E. LaBerge. 1978. *The Nest Biology of Bees* Andrena (Melandrena) regularis *Malloch and* Andrena (Melandrena) carlini *Cockerell (Hymenoptera: Andrenidae)*. State of Illinois Natural History Survey Division, Biological Note no. 108.

Sherwood, Dave. 2014. Caterpillar clash: The budworm returns. *Northern Woodlands* Winter: 28. Available via northernwoodlands.org/pdf/NW_Winter2014.pdf.

Taylor, Eric L., A. Gordon Holley, and Melanie Kirk. 2007. *Pesticide Development a Brief Look at the History*. Southern Regional Extension Forestry SREF-FM-010.

Tengo, Jan, and Gunnar Bergstrom. 1976. Comparative analyses of lemon-smelling secretions from heads of *Andrena* F. (Hymenoptera, Apoidea) bees. *Comparative Biochemistry and Physiology* 55(2): 179–188.

Thaler, G. R., and R. C. Plowright. 1980. The effect of aerial insecticide spraying for spruce budworm control on the fecundity of entomophilous plants in New Brunswick. *Canadian Journal of Botany* 58(18): 2022–2027.

U.S. Department of Agriculture. 2012 Census of Agriculture, Specialty Crops. agcensus.usda.gov/Publications/2012/Online_Resources/Specialty_Crops/SCROPS.pdf.

U.S. Environmental Protection Agency, OCSPP. How we assess risks to pollinators. epa.gov/pollinator-protection/how-we-assess-risks-pollinators.

Wilson, Joseph, and Olivia Messinger Carril. 2015. *The Bees in Your Backyard: A Guide to North America's Bees*. Princeton, NJ: Princeton University Press.

Cinderella *Ceratina* and Bees Down on the Farm

Blitzer, Eleanor J., Jason Gibbs, Mia G. Park, and Bryan N. Danforth. 2016. Pollination services for apples are dependent on diverse wild bee communities. *Agriculture, Ecosystems and Environment* 221: 1–7.

Chagnon, M., J. Gingras, and D. De Oliveira. 1993. Complementary aspects of strawberry pollination by honey and indigenous bees (Hymenoptera). *Journal of Economic Entomology* 86(2): 416–420.

Dimitri, Carolyn, Anne Effland, and Neilson Conklin. 2005. *The 20th Century Transformation of U.S. Agriculture and Farm Policy*. Bulletin of the U.S. Department of Agriculture Economic Research Service, no. 3. Available via ers.usda.gov/webdocs/publications/eib3/13566_eib3_1_.pdf.

Garibaldi, Lucas A., Ingolf Steffan-Dewenter, Rachael Winfree, et al. 2013. Wild pollinators enhance fruit set of crops regardless of honey bee abundance. *Science* 339(6127): 1608–1611.

Jensen, Carol A. 2008. *Images of America: Brentwood, California*. Mt. Pleasant, SC: Arcadia Publishing.

Klatt, Björn K., Andrea Holzschuh, Catrin Westphal, Yann Clough, Inga Smit, Elke Pawelzik, and Teja Tscharntke. 2014. Bee pollination improves crop quality, shelf life and commercial value. *Proceedings of the Royal Society of London B: Biological Sciences* 281(1775): 20132440.

Klein, Alexandra-Maria, Bernard E. Vaissière, James H. Cane, Ingolf Steffan-Dewenter, Saul A. Cunningham, Claire Kremen, and Teja Tscharntke. 2007. Importance of pollinators in changing landscapes for world crops. *Proceedings of the Royal Society of London B: Biological Sciences* 274(1608): 303–313.

Kremen, Claire, Neal M. Williams, and Robbin W. Thorp. 2002. Crop pollination from native bees at risk from agricultural intensification. *Proceedings of the National Academy of Sciences USA* 99(26): 16812–16816.

Oertel, E., E. C. Martin, and N. P. Nye. 1980. *Beekeeping in the United States*. Agricultural Handbook no. 335. Washington, DC: U.S. Department of Agriculture.

Rau, P. 1928. The nesting habits of the little carpenter bee, *Ceratina calcarata*. *Annals of the Entomological Society of America* 21(3): 380–397.

Rehan, Sandra M. 2011. Evolutionary origin and maintenance of sociality in the small carpenter bees. Ph.D. diss., Brock University.

Rehan, Sandra M., and Miriam H. Richards. 2010. Nesting biology and subsociality in *Ceratina calcarata* (Hymenoptera: Apidae). *The Canadian Entomologist* 142(1): 65–74.

Sampson, Blair J., Robert G. Danka, and Stephen J. Stringer. 2004. Nectar robbery by bees *Xylocopa virginica* and *Apis mellifera* contributes to the pollination of rabbiteye blueberry. *Journal of Economic Entomology* 97(3): 735–740.

Starkman, Naomi. 2009. Farmland conservation: The important lesson of

Brentwood, California. *Civil Eats*, August 10. Available via civileats.com/2009/08/10/farmland-conservation-the-important-lesson-of-brentwood-california.

University of California Berkeley Urban Bee Lab. Best bee plants for California. helpabee.org/best-bee-plants-for-california.html.

Vanderplanck, Maryse, Romain Moerman, Pierre Rasmont, Georges Lognay, Bernard Wathelet, Ruddy Wattiez, and Denis Michez. 2014. How does pollen chemistry impact development and feeding behaviour of polylectic bees? *PLoS One* 9(1): e86209.

Life, Death, and Thievery in the Dark

Burkle, Laura A., John C. Marlin, and Tiffany M. Knight. 2013. Plant-pollinator interactions over 120 years: Loss of species, co-occurrence, and function. *Science* 339: 1611–1615.

Danforth, Bryan N. 1989. Nesting behavior of four species of *Perdita* (Hymenoptera: Andrenidae). *Journal of the Kansas Entomological Society* 62(1): 59–79.

Eickwort, George C. 1981. Aspects of the nesting biology of five Nearctic species of *Agapostemon* (Hymenoptera: Halictidae). *Journal of the Kansas Entomological Society* 54(2): 337–351.

Frankie, Gordon, Robbin Thorp, Rollin Coville, and Barbara Ertter. 2014. *California Bees and Blooms: A Guide for Gardeners and Naturalists.* Berkeley, CA: Heyday.

Lutz, Diana. 2013. Walking in the footsteps of 19th- and 20th-century naturalists, scientists find battered plant-pollinator network. *The Source*, February 28. Available via source.wustl.edu/2013/02/walking-in-the-footsteps-of-19th-and-20thcentury-naturalists-scientists-find-battered-plantpollinator-network.

Maryland Department of Natural Resources. Common Maryland bees. dnr2.maryland.gov/wildlife/Documents/CommonBees.pdf.

Michener, Charles D. 1979. Biogeography of the bees. *Annals of the Missouri Botanical Garden* 66(3): 277–347.

Michener, Charles D. 2007. *The Bees of the World.* 2nd ed. Baltimore, MD: Johns Hopkins University Press.

Michener, Charles D., Ronald J. McGinley, and Bryan N. Danforth. 1994. *The Bee Genera of North and Central America (Hymenoptera: Apoidea)*. Washington, DC: Smithsonian Institution Press.

Minckley, Robert L., and John S. Ascher. 2012. Preliminary survey of bee (Hymenoptera: Anthophila) richness in the northwestern Chihuahuan Desert. In G. J. Gottfried, P. F. Ffolliott, B. S. Gebow, L. G. Eskew, and L. C. Collins, eds. *Merging Science and Management in a Rapidly Changing World: Biodiversity and Management of the Madrean Archipelago III*. Fort Collins, CO: U.S. Department of Agriculture, Forest Service, Rocky Mountain Research Station, 138–143.

Pascarella, John B. Bees of Florida. entnemdept.ufl.edu/HallG/Melitto/Intro.htm.

Pollinator Health Task Force. National strategy to promote the health of honey bees and other pollinators. whitehouse.gov/sites/default/files/microsites/ostp/Pollinator%20Health%20Strategy%202015.pdf.

Rozen, Jerome G. 1965. Biological notes on the cuckoo bee genera *Holcopasites* and *Neolarra* (Hymenoptera: Apoidea). *Journal of the New York Entomological Society* 73(2): 87–91.

Rozen, Jerome G. 1992. Systematics and host relationships of the cuckoo bee genus *Oreopasites* (Hymenoptera: Anthophoridae: Nomidinae). *American Museum Novitates* 3046: 56.

Rozen, Jerome George, and H. Glenn Hall. 2011. Nesting and developmental biology of the cleptoparasitic bee *Stelis ater* (Anthidiini) and its host, *Osmia chalybea* (Osmiini) (Hymenoptera, Megachilidae). *American Museum Novitates* no. 3707. Available via digitallibrary.amnh.org/handle/2246/6101.

Rozen, Jerome G., and Soliman M. Kamel. 2008. Hospicidal behavior of the cleptoparasitic bee *Coelioxys (Allocoelioxys) coturnix*, including descriptions of its larval instars (Hymenoptera: Megachilidae). *American Museum Novitates* 3636(1): 1.

Rozen, Jerome G., and Barbara Rozen. 1986. Bionomics of crepuscular bees associated with the plant *Psorothamnus scoparius* (Hymenoptera: Apoidea). *Journal of the New York Entomological Society* 94(4): 472–479.

Rozen, Jerome George, Kathleen R. Eickwort, and George C. Eickwort. 1978.

The bionomics and immature stages of the cleptoparasitic bee genus *Protepeolus* (Anthophoridae, Nomadinae). *American Museum Novitates* no. 2640. Available via digitallibrary.amnh.org/handle/2246/2953.

Sheffield, Cory S., Alana Pindar, Laurence Packer, and Peter G. Kevan. 2013. The potential of cleptoparasitic bees as indicator taxa for assessing bee communities. *Apidologie* 44(5): 501–510.

Skroch, Matt. Sky islands of North America: A globally unique and threatened inland archipelago. terrain.org/articles/21/skroch.htm.

Warshall, Peter. n.d. *The Madrean Sky Island Archipelago: A Planetary Overview*. U.S. Forest Service. Available via fs.fed.us/rm/pubs_rm/rm_gtr264/rm_gtr264_006_018.pdf.

Bees in the Grass: Rethinking Normal

Crownover, Matt. How many acres are needed for an 18 hole golf course? golftips.golfsmith.com/many-acres-needed-18-hole-golf-course-1812.html.

Dobbs, Emily K. 2013. Enhancing beneficial insect biodiversity and biological control in turf: Mowing height, naturalized roughs, and Operation Pollinator. Master's thesis, University of Kentucky.

Dobbs, Emily K., and Daniel A. Potter. 2015. Forging natural links with golf courses for pollinator-related conservation, outreach, teaching, and research. *American Entomologist* 61(2): 116–123.

Dyer, Judith G., and Alvin E. Shinn. 1978. Pollen collected by *Calliopsis andreniformis* Smith in North America (Hymenoptera: Andrenidae). *Journal of the Kansas Entomological Society* 51(4): 788–795.

Eickwort, George C., and Howard S. Ginsberg. 1980. Foraging and mating behavior in Apoidea. *Annual Review of Entomology* 25: 421–446.

Gerling, Dan, and Henry R. Hermann. 1978. Biology and mating behavior of *Xylocopa virginica* L. (Hymenoptera, Anthophoridae). *Behavioral Ecology and Sociobiology* 3(2): 99–111.

Goulson, Dave, Jason W. Chapman, and William O. H. Hughes. 2001. Discrimination of unrewarding flowers by bees: direct detection of rewards and use of repellent scent marks. *Journal of Insect Behavior* 14(5): 669–678.

Grube, Arthur, David Donaldson, Timothy Kiely, and La Wu. 2011. *Pesticide Industry Sales and Usage 2006 and 2007 Market Estimates*. Washington, DC: U.S. Environmental Protection Agency.

The Hall & Duck Trust. History of the lawnmower. Part 1: 1830–1850s. hdtrust.co.uk/hist01.htm.

Jabr, Ferris. 2013. Outgrowing the traditional grass lawn. *Scientific American: Brainwaves*, July 29. blogs.scientificamerican.com/brainwaves/outgrowing-the-traditional-grass-lawn.

Kukuk, Penelope. 1985. Evidence for an antiaphrodisiac in the sweat bee *Lasioglossum (Dialictus) zephyrum*. *Science* 227(4687): 656–657.

The R & A. 2015 Golf around the World 2015. Available via randa.org/TheRandA/AboutTheRandA/DownloadsAndPublications.

Rau, Phil. 1933. *The Jungle Bees and Wasps of Barro Colorado Island*. St. Louis, MO: Von Hoffman Press.

Richards, Miriam H., and Laurence Packer. 1994. Trophic aspects of caste determination in *Halictus ligatus*, a primitively eusocial sweat bee. *Behavioral Ecology and Sociobiology* 34: 385–391.

Richards, Miriam H., and Laurence Packer. 1998. Demography and relatedness in multiple-foundress nests of the social sweat bee, *Halictus ligatus*. *Insectes Sociaux* 45(1): 97–109.

Scott Jenkins, Virginia. 1994. *The Lawn: A History of an American Obsession*. Washington, DC: Smithsonian Books.

Shepherd, Matthew. 2002. *Making Room for Native Pollinators: How to Create Habitat for Pollinator Insects on Golf Courses*. Portland, OR: The U.S. Golf Association and the Xerces Society.

Smith, Lionel S., and Mark D. E. Fellowes. 2013. Toward a lawn without grass: the journey of the imperfect lawn and its analogues. *Studies in the History of Gardens and Designed Landscapes* 33(3): 157–169.

U.S. Environmental Protection Agency. Outdoor water use in the United States. www3.epa.gov/watersense/pubs/outdoor.html.

Wilson, Joseph, and Olivia Messinger Carril. 2015. *The Bees in Your Backyard: A Guide to North America's Bees*. Princeton, NJ: Princeton University Press.

Citizen Science and the Great Sunflower Project

Berenbaum, May, Peter Bernhardt, Stephen Buchmann, et al. 2007. *Status of Pollinators in North America*. Washington, DC: National Academies Press.

The Great Sunflower Project. greatsunflower.org/node/1?no_cache = 1477116934.

Hurd, Paul D., Wallace E. LaBerge, and E. Gorton Linsley. 1980. *Principle Sunflower Bees of North America with Emphasis on the Southwestern United States (Hymenoptera: Apoidea)*. Smithsonian Contributions to Zoology no. 310. Washington, DC: Smithsonian Institution.

Neff, J. L., B. B. Simpson, and L. J. Dorr. 1982. The nesting biology of *Diadasia afflicta* Cress. (Hymenoptera: Anthophoridae). *Journal of the Kansas Entomological Society* 55(3): 499–518.

Parker, Frank D. 1981. How efficient are bees in pollinating sunflowers? *Journal of the Kansas Entomological Society* 54(1): 61–67.

Rau, Phil. 1938. Additional observations on the sleep of insects. *Annals of the Entomological Society of America* 31(4): 540–556.

Rau, Phil, and Nellie Rau. 1916. The sleep of insects: An ecological study. *Annals of the Entomological Society of America* 9(3): 227–274.

Rozen, Jerome G., and Hadel H. Go. 2015. Descriptions of the egg and mature larva of the bee *Chelostoma (Prochelostoma) philadelphi* with additional notes on nesting biology (Hymenoptera: Megachilidae: Megachilinae: Osmiini). *American Museum Novitates* 3844: 7.

Thorp, Robbin, and Dennis L. Briggs. 1980. Bees collecting pollen from other bees (Hymenoptera: Apoidea). *Journal of the Kansas Entomological Society* 53(1): 166–170.

Wcislo, William T., and James H. Cane. 1996. Floral resource utilization by solitary bees (Hymenoptera: Apoidea) and exploitation of their stored foods by natural enemies. *Annual Review of Entomology* 41: 257–286.

Xerces Society for Invertebrate Conservation. Native bee pollination of hybrid sunflowers. xerces.org/wp-content/uploads/2008/10/factsheet_sunflower_pollination.pdf.

Zuparko, Robert L. The published names of TDA Cockerell. essig.berkeley.edu/documents/tda_cockerell/Cockerell_Intro.pdf.

Further Reading

Buchmann, Stephen L., and Gary Paul Nabhan. 1997. *The Forgotten Pollinators*. Washington, DC: Island Press.

This is the original book on native bees.

Droege, Sam, and Laurence Packer. 2015. *Bees: An Up-Close Look at Pollinators Around the World*. Minneapolis, MN: Voyageur Press.

A coffee table book of beautiful bees with interesting tidbits added in.

Frankie, Gordon, Robbin Thorp, Rollin Coville, and Barbara Ertter. 2014. *California Bees and Blooms: A Guide for Gardeners and Naturalists*. Berkeley, CA: Heyday.

A useful overview of bees and how to garden for them, even if you don't live in California.

Frey, Kate, and Gretchen LeBuhn. 2016. *The Bee-Friendly Garden: Design an Abundant Flower-filled Yard that Nurtures Bees and Supports Biodiversity*. Berkeley, CA: Ten Speed Press.

Written by a garden designer and a sometime instructor at the Arizona Bee Course, who also started the Great Sunflower Project.

Goulson, Dave. 2014. *A Sting in the Tale: My Adventures with Bumblebees*. London: Picador.

An armchair read by a British entomologist.

Hunter, Dave, and Jill Lightner. 2016. *Mason Bee Revolution: How the Hardest Working Bee Can Save the World One Backyard at a Time*. Seattle: Skipstone.

A more detailed look at BOBs and Dave Hunter's grand hopes for them.

Packer, Laurence. 2014. *Keeping the Bees: Why All Bees Are at Risk and What We Can Do to Save Them.* New York: HarperCollins.
Another armchair read written by one of the instructors at the Arizona Bee Course.

Williams, Paul, Robbin Thorp, Leif Richardson, and Sheila Colla. 2014. *Bumble Bees of North America: An Identification Guide.* Princeton, NJ: Princeton University Press.
Bumble bees are big and showy and therefore noticeable. This smallish book is portable if you want to do some bumble bee ID outside your home town.

Wilson, Joseph, and Olivia Messinger Carril. 2015. *The Bees in Your Backyard: A Guide to North America's Bees.* Princeton, NJ: Princeton University Press.
This book has lots of fun facts, photos, and identification tips.

The Xerces Society and Marla Spivak. 2011. *Attracting Native Pollinators: Protecting North America's Bees and Butterflies.* North Adams, MA: Storey.
Comprehensive and accessible, a great first read on pollinators.

Websites

Bee Basics: An Introduction to Our Native Bees
www.fs.usda.gov/Internet/FSE_DOCUMENTS/stelprdb5306468.pdf

Clay Bolt's photo website
claybolt.photoshelter.com/index

Native Bee Conservancy
nativebeeconservancy.org

Penn State Garden Certification Program
ento.psu.edu/pollinators/public-outreach/cert

Pollinator Partnership
pollinator.org

Rollin Coville's photo website

covillephotos.com

University of California Berkeley Urban Bee Lab

helpabee.org

U.S. Geological Society Bee Inventory and Monitoring Lab photos

flickr.com/photos/usgsbiml

The Xerces Society for Invertebrate Conservation

xerces.org

Citizen Science and Interactive Sites

Bug Guide

bugguide.net/node/view/59

Bee photos posted here may also be identified for you.

Bumble Bee Watch

bumblebeewatch.org

A nationwide citizen science project.

Discover Life

discoverlife.org/mp/20q?search = Apoidea

Interactive key to bees and a place to post photos and potentially get them identified.

The Great Sunflower Project

greatsunflower.org

This is the website for Gretchen LeBuhn's massive citizen science project. It also has a compilation of some of the data that have been collected.

Acknowledgments

Writing a bee book as a bee novice has been both a joy and a terror. The joy comes from the learning and the people I met. Writing this book as a series of stories has allowed me to follow my whim of interest down whatever paths looked interesting. Along those paths I've had the great pleasure of meeting an astonishing number of helpful, knowledgeable, passionate people. The terror comes from a fear of making mistakes. I've read books, waded through papers, talked to all sorts of people, and tried to put it all together into true stories about bees and the people who love them. Many have fact-checked for me, and all errors that remain are mine alone.

I would like to thank all the people that have helped make this happen. First there are all the people who took time out of their lives to be interviewed, often taking me out into the field or farm or into the lab: Gordon Frankie, Jerry Rozen, Gordon Wardell, Gene Brandi, Frank Drummond, Mark Hoban, Jim Cane, Al Courchesne, Linnea Beedy, Dave Hunter, Kris Fade, Bernie Mach, Dan Potter, Eric Mussen, Gretchen LeBuhn, Gail Langellotto-Rhodaback, Theresa Gaffney, Terry Griswold, Rich Bireley, Jason Whitecliff, Emily Dobbs, Ian Lane, Mary Meyer, Scott Bender, David Tarpy, Neal Williams, May Berenbaum, Elina Nino, Wayne Wehling, the folks at the Bee Biodiversity Initiative, Sandra Rehan, Walt Osborne, Matt Shaffer, Connie Schmotzer, and Lydia Martin. Special thanks go to Robbin Thorp, who has patiently answered so many questions for me.

Thanks to the photographers who provided all the beautiful bee photos, particularly Clay Bolt, Rollin Coville, and Sam Droege and the folks at the U.S. Geological Survey lab who were my go-to people when I needed a photo.

The book wouldn't be what it is without my writing group—thanks to all of you. Thanks to Jessica Murphy Moo, who taught me the importance of writing scenes in nonfiction.

Elizabeth Wales, my agent, bravely took on an author with none of the credentials a writer is supposed to have nowadays. She guided me through the process and talked me off the ledge a few times. Thanks to my editor, Tom Fischer, and the folks at Timber Press who helped get me through the book-making process.

My children accepted with grace and an extraordinary amount of grief (bees in the fridge!) all the time I took to work on this book. Lastly, there is my husband for whom mere thanks are not enough. He supported the idea of me writing a bee book from the beginning and never once said he thought I was nuts for trying to do it. He let me use up our airline miles and run away to play in bee fields while he stayed home and looked after the kids. He listened to me go on and on about bees when we went out to dinner. He reviewed chapters, parts of chapters, and sometimes even sentences. Without his support, this book would not exist.

Photo and Illustration Credits

Clay Bolt, pages 14, 16 (top), 33, 43, 55, 57, 66, 67 (bottom), 72, 90, 93, 146, 156 (bottom), 164, 175, 177, 180

Alan Bryan, pages 13, 18

Ellen Bulger, page 89

Rollin Coville, pages 50, 52, 67 (top), 74, 75, 77, 85, 108, 109, 113 (top), 117, 124, 125, 136 (top), 139 (top), 143, 145, 147 (bottom), 157 (bottom), 159 (bottom), 168, 170, 178, 179 (top), 182, 184 (top)

Sam Droege's USGS Bee Inventory and Monitoring Lab, pages 8, 10, 15, 16 (bottom), 27, 56, 68, 86, 87, 91, 113 (bottom), 119, 120, 121, 130 (bottom), 133, 139 (bottom), 142, 147 (top), 149, 156 (top), 157 (top), 159 (top), 169, 179 (bottom), 184 (bottom), 192, 194

Bruce Lund, page 187

Randy Oliver, ScientificBeekeeping.com, page 31

Lynette Elliott, page 63

Paul H. Williams et al., *Bumble Bees of North America*, Princeton University Press, 2014, page 48

Wikimedia

Used under a Creative Commons Attribution 3.0 Unported license

Gideon Pisanty (Gidip) גדעון פיזנטי, page 35

Used under a Creative Commons Attribution-Share Alike 2.5 Generic license

Nino Barbieri, page 25

All other photos are by the author.

Index